Statistical Methods in Longitudinal Research

Volume II

Time Series and
Categorical Longitudinal Data

Statistical Methods in Longitudinal Research

Volume II

Time Series and Categorical Longitudinal Data

Edited by

Alexander von Eye

Department of Human Development and Family Studies
The Pennsylvania State University
University Park, Pennsylvania

ACADEMIC PRESS, INC.
Harcourt Brace Jovanovich, Publishers

Boston   San Diego   New York
London   Sydney   Tokyo   Toronto

ACADEMIC PRESS, INC.
1250 Sixth Avenue, San Diego, CA 92101

United Kingdom Edition published by
ACADEMIC PRESS LIMITED
24–28 Oval Road, London NW1 7DX

Library of Congress Cataloging-in-Publication Data

Statistical methods in longitudinal research/edited by Alexander von
    Eye.
        p.   cm.—(Statistical modeling and decision science)
    Bibliography: p.
    Includes index.
    Contents: v. 1. Principles and structuring change.—v. 2. Time
series and categorical longitudinal data.
    ISBN 0-12-724960-5 (v. 1: alk. paper).—ISBN 0-12-724961-3 (v.
2: alk. paper) ISBN 0-12-724962-1 (v. 1; pbk).—ISBN 0-12-724963-X (v. 2; pbk)
    1. Social sciences—Research—Statistical methods.   2. Social
sciences—Research—Longitudinal studies.   I. Eye, Alexander von.
    II. Series
    HA29.S7835   1990
    519.5—dc20
                                                            89-32965
                                                            CIP

Printed in the United States of America
90 91 92 93   9 8 7 6 5 4 3 2 1

Contents of Volume II

Contents of Volume I

Contributors

Numbers in parentheses refer to the pages on which the authors' contributions begin. Numbers in italics indicate contributions to Volume II.

Mark S. Aber (151), *Department of Psychology, The University of Illinois, Champaign, Illinois 61820*

R. Darrell Bock (*289*), *Departments of Education and Behavioral Science, The University of Chicago, Chicago, Illinois 60637*

Jeffrey A. Burr (3), *Department of Sociology, State University of New York at Buffalo, 430 Park Hall, Buffalo, New York 14260*

Clifford C. Clogg (*409*), *Departments of Sociolocy and Statistics, and Population Issues Research Center, The Pennsylvania State University, University Park, Pennsylvania 16802*

Mary Delaney (35), *Department of Human Development and Family Studies, The Pennsylvania State University, University Park, Pennsylvania 16802*

Scott R. Eliason (*409*), *Department of Sociology, The Pennsylvania State University, University Park, Pennsylvania 16802*

Edgar Erdfelder (*471*), *Psychologisches Institut, Universität Bonn, D-5300 Bonn 1, Federal Republic of Germany*

Paul A. Games (81), *Department of Human Development and Family Studies, Pennsylvania State University, University Park, Pennsylvania 16802*

John M. Grego (*409*), *Department of Statistics, University of South Carolina, Columbia, South Carolina 29208*

Randy J. Larsen (*319*), *Department of Psychology, The University of Michigan, Ann Arbor, Michigan 48109–1346*

J. J. McArdle (151), *Department of Psychology, The University of Virginia, Charlottesville, Virginia 22901*

William Meredith (125, 387), *Department of Psychology, University of California at Berkeley, Berkeley, California 94720*

John R. Nesselroade (3), *Department of Human Development and Family Studies, The Pennsylvania State University, University Park, Pennsylvania 16802*

Trond Petersen (259), *Walter A. Haas School of Business, Barrows Hall 350, University of California at Berkeley, Berkeley, California 94720*

David Rindskopf (443), *Graduate School and University Center, City University of New York, New York, New York 10036–8099*

Michael J. Rovine (35), *Department of Human Development and Family Studies, The Pennsylvania State University, University Park, Pennsylvania 16802*

Bernhard Schmitz (351), *Max-Planck-Institut für Bildungsforschung, Lentzeallee 94, D-1000 Berlin 33, Federal Republic of Germany*

Kathryn A. Szabat (511), *Department of Management, LaSalle University, Philadelphia, Pennsylvania 19141*

David Thissen (289), *Department of Psychology, University of Kansas, Lawrence, Kansas 66045*

John Tisak (125, 387), *Department of Psychology, Bowling Green State University, Bowling Green, Ohio, 43402*

Alexander von Eye (545), *Department of Human Development and Family Studies, The Pennsylvania State University, University Park, Pennsylvania 16802*

Phillip Wood (225), *Psychology Department, University of Missouri-Columbia, Columbia, Missouri 65211*

Preface

Longitudinal investigations are crucial to the study of development and change. In particular, intraindividual change can be depicted only via repeated observations. To many social scientists, the statistical analysis of repeated measurement data is a challenge because of problems with dependent measures, loss of subjects, availability of computer expertise, or sample size. In many instances, researchers have problems matching substantive questions and statistical methods.

In the methodological literature, considerable progress has been made in the discussion of methods for handling longitudinal data. This progress has remained unnoticed by many, and teaching as well as application of statistical methods has not benefitted from the advances in methodology. In other words, there is a gap between the state of the art of methodology and the application of statistics. It is the main goal of these volumes to narrow this gap.

To illustrate the progress that has been made in the development of statistical methodology, each chapter presents new aspects of methodology or statistics. Here, the attribute "new" applies to all facets of the methods for analysis of longitudinal data. For instance, "new" can mean that methods well-known in certain fields of application are applied to a problem from another field for the first time; it can mean that problems with the application of a software package are identified for the first time; it can mean that a statistical method is further developed to accommodate an extended range of application; it can mean that characteristics of a method are outlined for the first time, so that applicants have a clearer picture of when best to apply this method; it can mean that well-known problems are expressed in terms never used before to express them, so that the researcher has a language available that helps structure problem

xi

specification. All chapters in these volumes contribute something new in this sense. None is merely a write-up of a well-known approach.

A second goal of these volumes concerns the application of the new methods. Reasons for the wide gap between the results obtained in the development of statistical methods and their application in empirical research include the lack of computer software and the lack of instructions on how to use available software in applying new methods. This book emphasizes computational statistics. Each chapter that explicitly discusses statistical methods and their application contains instructions and examples of how to use a particular program or identifies available programs. In addition, a companion book by Rovine and von Eye will be published by Academic Press in which examples of program applications are detailed.

The targeted readership of these books includes students of development and change. The books are not oriented toward a particular discipline. Rather, the scope is broad enough to include researchers from all empirical sciences, including the social sciences, economics, biology, or medicine.

The volumes contain 16 chapters, grouped in four sections. This volume contains Sections 1 and 2, and Volume II contains Sections 3 and 4. The first section covers problems of general interest. It begins with a discussion of change processes (Burr and Nesselroade). Here, central terms are explicated, and general problems are specified. The second chapter discusses the problem of missing data, which is almost ubiquitous in longitudinal research in the social sciences (Rovine and Delaney). Approaches to repeated measurement analysis of variance with covariates are discussed in tandem with problems of the application of commercial software in Chapter 3 (Games).

The second section includes chapters on the structuring of change. It begins with a chapter on longitudinal factor analysis (Tisak and Meredith). This approach has only recently been "redetected," drawing renewed attention from both methodologists and applicants. An approach that has changed the perspectives in the domain of longitudinal research is structural equation modeling. The next chapter covers this approach (McArdle and Aber). Rarely applied in present longitudinal research, but nevertheless promising, are methods of scaling. The chapter by Wood introduces readers to this methodology.

Volume II begins with the third section, which covers the analysis of time series. It opens with a discussion of event history analysis (Petersen), a method that, thus far, has found most application in sociology but is of great interest to other fields as well. The next chapter treats growth curve analysis (Thissen and Bock), adopting the perspective of mathematical modeling of observed processes. The following chapter, by Larsen, covers spectral analysis. It presents methods of decomposing time series into elementary functions, trigonometric functions in this case. Each of these functions can be interpreted in terms of assumed processes. The fourth chapter in this section discusses time series analysis within the Box and Jenkins framework (Schmitz), including sections on multivariate analysis. The last chapter of this section introduces the reader to methods of segmenting multivariate response curves within the framework of the general linear model (Tisak and Meredith). In recognition of the contribution of Ledyard R Tucker to this methodology, the authors use the term the term "tuckerizing curves."

The fourth section discusses developments in the analysis of repeatedly observed categorical data. It begins with a new approach to formulating longitudinal models for log-linear modeling (Clogg, Eliason, and Grego). The authors adopt the design matrix approach to specifying longitudinal models. Closely related to log-linear modeling is latent class analysis. This approach is covered in the second chapter of this section by Rindskopf, which includes a section on how to use IMSL programs to do parameter estimations. In the third chapter, Erdfelder covers finite mixture distributions. This approach allows one to test assumptions concerning univariate as well as multivariate discrete or continuous distributions. Custom-tailored hypothesis testing in contingency tables can be done with prediction analysis (Szabat), as shown in the fourth chapter, which includes sections describing characteristics of the statistical tests applied to evaluate the model. The final chapter discusses exploratory configural frequency analysis (von Eye). This method allows one to test cellwise whether assumptions concerning longitudinal processes are fulfilled.

I am indebted to a plethora of individuals who supported this work. Viewed from a longitudinal perspective, I would like first to thank J. R. Nesselroade, who encouraged me and helped me to get started. I also would like to thank Academic Press, and in particular Klaus Peters, who was very interested and supportive. Many friends supported this enterprise by reviewing chapters: Constance Jones, Gustav A. Lienert, Jack McArdle, John Nesselroade, Mike Rovine, Holger Wessels, and Phil Wood. I thank them for their efforts and their wisdom. For her invaluable secretarial support, I wish to thank Nancy Cole. I am deeply indebted to the authors who contributed outstanding chapters, often under adverse conditions: One author's computer exploded while he suffered through the drafting process in Hawaii, but he nevertheless submitted a very impressive paper. All authors responded in a very professional way to requests for revisions. Some authors' patience was tried because other authors needed time to complete their chapters. I thank them all very much. Without their efforts, these volumes would not have been possible. Most of all, I would like to thank Donata, Maxine, Valerie, and Julian, who provide the right context and a longitudinal perspective for mutual development in our family.

Alexander von Eye

III Analysis of Time Series

Chapter 7

Analyzing Event Histories*

TROND PETERSEN

Walter A. Haas School of Business
University of California, Berkeley
Berkeley, California

Abstract

Event histories are generated by so-called failure time processes and take this form. The dependent variable—for example, some social state—is discrete or continuous. Over time it evolves as follows. For finite periods of time (that is, from one calendar date to another) it stays constant at a given value. At a later date, which is a random variable, the dependent variable jumps to a new value. The process evolves in this manner from the calendar date, when one change occurs, to a later date, when another change occurs. Between the dates of the changes, the dependent variables stays constant.

Data on such processes typically contain information about (a) the date a sample member entered a social state, (b) the date the state later was left, if left, and (c) the value of the next state entered, and so on.

In the analysis of such data the foci are on what determines the amount of time spent in each state and on what determines the value of the next state entered. This chapter describes how one can use continuous-time

* This chapter was commissioned by the National Institute of Consumer Research in Norway, who provided support for the research. I also acknowledge gratefully the financial support from the National Institute of Aging, Grant #AG04367. I am grateful to Jim Duke and Karl Eschbach for research assistance, and I thank the editor, Alexander von Eye, for editorial comments.

259

hazard rate models to address these two foci when analyzing event histories.

1 Introduction

Event histories are generated by so-called failure time processes and take this form. The dependent variable—for example, some social state—is discrete or continuous. Over time it evolves as follows. For finite periods of time (that is, from one calendar date to another) it stays constant at a given value. At a later date, which is a random variable, the dependent variable jumps to a new value. The process evolves in this manner from the calendar date, when one change occurs, to a later date, when another change occurs. Between the dates of the changes, the dependent variable stays constant.

Data on such processes typically contain information about the date a sample member entered a social state, for example an employment state, the date the state later was left, if left, and the value of the next state entered, and so on.

In the analysis of such data the foci are on what determines the amount of time spent in each state and on what determines the value of the next state entered. Typically one would like to assess the effects of covariates on the amount of time spent in a state and on the value of the next state entered.

This chapter discusses three types of failure time or jump processes. The simplest type obtains when there is a single state that can be occupied only once. This type is referred to as a *single-state nonrepeatable event process*. A person currently in the state may or may not leave it. If it is left, one does not distinguish between different reasons for leaving or different destination states. The state, once left, cannot be reentered. An example is the process of entry into first marriage, provided one makes no distinction between religious and secular marriages (see, e.g., Hernes, 1972). Another example is mortality (see, e.g., Vaupel, Manton, and Stallard, 1979). Being alive is a state that cannot be reentered and typically one does not distinguish between different destination states, heaven or hell.

The second type of process considered is the *multistate process*. The state currently occupied can be left for several distinct reasons. For example, in AIDS research, an AIDS victim may die from Kaposi's sarcoma, pneumonia, or some other complication. One would be interested in detecting which complication is most deadly, which in turn would allow one to make judgments about the most effective alleviating measures. In most cases the number of states is finite, that is, the state-space is discrete, but in some instances it is continuous, as, for example, in the analysis of individual earning histories.

Finally, I consider *repeatable-event processes*. In these processes a person can occupy a state several times. Job histories fall within this class of process (see Tuma, 1976). The researcher focuses on the amount of time spent in each job. Each sample member may contribute more than one job. Typically, such processes also have multiple states: employed, unemployed, and out of the labor force (see Flinn and Heckman, 1983).

In all three types of failure time processes the objective of the empirical analysis is, as stated before, to analyze the determinants of the amount of time that elapses between changes and the value of the destination state once a change occurs.[1]

The remainder of the chapter is organized in nine sections. Section 2 outlines the basic strategy for analyzing event histories by means of hazard rate models. Section 3 explains how explanatory variables can be introduced into the hazard rate. In Section 4 comparisons to more familiar regression-type models are made. Section 5 discussed repeatable-event processes. Section 6 discusses multistate processes, in both continuous and discrete state space. Section 7 addresses the issue of unobserved variables that influence the rate. Section 8 discusses briefly some theoretical models that may generate event histories. An empirical application is presented in Section 9. Section 10 concludes the chapter. Appendix A provides proofs of some of the results stated in Section 2, and Appendix B discusses some relevant computer programs.

2. Basic Concepts

2.1. Why Not Ordinary Regression Analysis?

Suppose the researcher has collected career histories on employees in a hierarchically organized company. The researcher might be interested in analyzing the determinants of promotion or the amount of time that elapses before a promotion occurs. Let t be the amount of time that elapsed before an employee was promoted or before he or she left the

---

[1] An exception to this characterization is the counting process framework (see Andersen and Borgan, 1985), where the concept of a failure time plays only a marginal role.

company without having been promoted, or before he or whe was last observed in the company without having received a promotion. Let x denote the vector of explanatory variables, for example, race, sex and marital status.

One may formulate a linear regression model as follows:

$$\ln t = \beta x + \varepsilon, \tag{1}$$

where $\ln t$ is the natural logarithm of t, β are the effect of parameters pertaining to x, and ε is a stochastic error term.

There are at least two problems with the approach in (1). *First*, it treats employees who were never promoted or who left the company without receiving a promotion in the same way as those who did experience a promotion. The former cases are referred to as *right-censored*. We know only that they had not experienced the event of interest when they were last observed in the company. A *second* problem with the formulation in (1) arises when the covariates in x change over time. The number of patterns that x may take over time can be very large, and to account for all of these on the right-hand side of (1) may be close to impossible.

One response to the first problem is to restrict analysis to those employees who were promoted. But this solution generates other problems. One of these is that there may be systematic differences between those who were promoted and those who were not. If the research interest is to assess the determinants of promotions, the bias introduced by excluding those who were not promoted may be severe. We will only learn about the amount of time that elapsed before a promotion among those who were promoted.

Another response to the problem of right-censoring would be to define a dummy variable C that is equal to 1 if a promotion occurred, and 0 otherwise, and then estimate a logit (or probit) model predicting the probability of having been promoted, as follows (in the logit case):

$$P[C = 1 \mid x] = \frac{\exp(\beta x)}{1 + \exp(\beta x)}. \tag{2}$$

The drawback of this procedure is that it ignores the amount of time that elapsed before a promotion or censoring occurred. Being promoted after six months is a different career trajectory from being promoted after six years, but (2) does not distinguish the two cases. Inserting t on the right-hand side of the equation would be erroneous, because one would then conflate a part of the dependent variable (i.e., t) with the independent variables. Also, (2) cannot account for time-varying covariates, unless one defines the probability in (2) separately for each observed time unit, say, each week or month (see Allison, 1982).

2.2. Solution to the Problem of Right-Censoring

The solution to the problems of right-censoring and time-dependent covariates is now described. Instead of focusing on the entire duration t, one proceeds in a more stepwise manner. The central idea is to divide the duration t into several segments. Let $t_k = t$ and set $t_0 = 0$. We then have k segments of time from duration 0 to duration t_k. The first segment covers the interval 0 to t_1, the second covers t_1 to t_2, and so on, up until t_{k-1} to t_k, where $0 = t_0 < t_1 < t_2 < \cdots < t_{k-1} < t_k = t$. Each segment has length $\Delta t = t_{j+1} - t_j$.

Now let T be the random variable denoting the amount of time spent in a state before a transition or censoring occurs. The hazard rate framework proceeds by specifying the probability that the state is left during the duration interval t_j to t_{j+1}, given that it was not left before t_j:

$$P[t_j \leq T < t_j + \Delta t \mid T \geq t_j], \qquad \text{where } t_j + \Delta t = t_{j+1}, \qquad (3)$$

and conversely, the probability that the state was not left in the duration interval t_j to t_{j+1}, given that it was not left before t_j, namely

$$P[T \geq t_{j+1} \mid T \geq t_j] = 1 - P[t_j \leq T < t_{j+1} \mid T \geq t_j], \quad t_{j+1} = t_j + \Delta t. \quad (4)$$

By means of these probabilities (3) and (4), defined for each small segment of time t_j to t_{j+1}, one can derive the probability that the state was not left before duration t_k as follows:

$$P(T \geq t_k) = \prod_{j=0}^{k-1} P(T \geq t_{j+1} \mid T \geq t_j), \quad t_0 = 0 \quad \text{and} \quad t_{j+1} = t_j + \Delta t, \quad (5)$$

which follows from rules for conditional probabilities.

Similarly, the probability that the state was left between duration t_k and $t_k + \Delta t$ follows as

$$P[t_k \leq T < t_k + \Delta t] = P(T \geq t_k)P(t_k \leq T < t_k + \Delta t \mid T \geq t_k)$$

$$= \prod_{j=0}^{k-1} P(T \geq t_{j+1} \mid T \geq t_j)P[t_k \leq T < t_k + \Delta t \mid T \geq t_k].$$

$$(6)$$

The interpretation of (5) is this: The probability of not having an event before duration t_k equals the probability of surviving beyond duration t_1 times the probability of surviving beyond duration t_2, given survival at t_1, and so on, up until the probability of surviving beyond duration t_k, given survival at t_{k-1}. The interpretation of (6) is this: The probability of leaving the state in the duration interval t_k to $t_k + \Delta t$ equals the probability of not leaving it before duration t_k, that is (5), times the probability of leaving the state between duration t_k and $t_k + \Delta t$, given that it was not left before duration t_k.

In conclusion, if one can specify the probability of a transition in a small time interval, that is, (3), given no transition prior to entry into the interval, then one can derive (5) and (6).

To complete the derivation of the hazard rate framework, one step remains. The choice of Δt, that is, the length of the time interval for which (3) is specified, is arbitrary. The convention, therefore, since time is continuous, is to let Δt approach zero, namely

$$\lim_{\Delta t \downarrow 0} P[t_j \leq T < t_j + \Delta t \mid T \geq t_j], \qquad \text{where } t_j + \Delta t = t_{j+1}. \tag{7}$$

Since time or duration T is an absolutely continuous variable, the probability of any specific realization of T is zero, and (7) is hence also equal to zero. As with other continuous variables, one therefore divides the probability in (3) with Δt, which yields a probability per time unit divided by the time unit itself. Then one takes the limit of this ratio as the time unit goes to zero. This operation yields the central concept in event-history analysis, the *hazard rate* (see Kalbfleisch and Prentice, 1980, p. 6)

$$\lambda(t_j) \equiv \lim_{\Delta t \downarrow 0} \frac{P[t_j \leq T < t_j + \Delta t \mid T \geq t_j]}{\Delta t} \tag{8}$$

which is a conditional density function: the density that the state is left at duration t_j, given that it was not left before duration t_j.

From (8) we get that for small Δt the probability of a transition, initially specified in (3), becomes

$$P[t_j \leq T < t_j + \Delta t \mid T \geq t_j] = \lambda(t_j)\Delta t, \qquad \text{when } \Delta t \text{ is small.} \tag{9}$$

Then inserting (9) into (4), we get

$$P[T \geq t_{j+1} \mid T \geq t_j] = 1 - \lambda(t_j)\Delta t, \qquad \text{where } t_{j+1} = t_j + \Delta t. \tag{10}$$

Inserting (10) into (5) yields

$$P[T \geq t_k] = \sum_{j=0}^{k-1} [1 - \lambda(t_j)\Delta t]; \tag{11}$$

then inserting (11) and (9) into (6) yields

$$P[t_k \leq T < t_k + \Delta t] = \sum_{j=0}^{k-1} [1 - \lambda(t_j)\Delta t]\lambda(t_k)\Delta t. \tag{12}$$

Equations (11) and (12) have the same interpretations as (5) and (6), but the right-hand sides are now expressed exclusively in terms of the hazard rate in (8).

Since time is absolutely continuous, the expressions in (11) and (12) must, as in (8), be evaluated as Δt goes to zero. When Δt goes to zero, the number of segments k goes to infinity, since $k = t_k/\Delta t$. Computing the limit of (11) as $\Delta t \downarrow 0$ and $k \to \infty$ yields the famous expression for the probability of surviving beyond duration t_k:

$$P[T \geq t_k] = \lim_{\substack{\Delta t \downarrow 0 \\ k \to \infty}} \prod_{j=0}^{k-1} [1 - \lambda(t_j)\Delta t]$$

$$= \exp\left[-\int_0^{t_k} \lambda(s)\,ds \right], \tag{13}$$

known as the *survivor function*. A proof of (13) is given in Appendix A. In the last equality of (13), s denotes duration in state.

Here I consider a proof of (13) in the special case when the hazard rate equals a constant θ for all t. The integral of $\lambda(t) = \theta$ from 0 to t_k then equals θt_k, and the survivor function in (13) is hence $\exp[-\theta t_k]$. In order to show this, using the limit operations on the right-hand side of the first equality in (13), set first $\Delta t = t_k/k$. Then let $k \to \infty$ (i.e., $\Delta t \downarrow 0$), which, for a fixed t_k, yields

$$P[T \geq t_k] = \lim_{k \to \infty} \prod_{j=0}^{k-1} \left[1 - \frac{\theta t_k}{k} \right]$$

$$= \lim_{k \to \infty} \left[1 - \frac{\theta t_k}{k} \right]^k$$

$$= \exp[-\theta t_k], \tag{14}$$

where the last equality follows from a well-known fact in calculus [see Apostol, 1967, equation (10.13), p. 380].

Finally, consider the limit of (12) as Δt goes to zero and k goes to infinity, but now first dividing by Δt on both sides of (12), for the same reason as in (8). We obtain the *density function* for the duration t_k as

$$f(t_k) \equiv \lim_{\Delta t \downarrow 0} \frac{P[t_k \leq T < t_k + \Delta t]}{\Delta t}$$

$$= \lim_{\substack{\Delta t \downarrow 0 \\ k \to \infty}} \prod_{j=0}^{k-1} [1 - \lambda(t_j)\Delta t] \frac{\lambda(t_k)\Delta t}{\Delta t}$$

$$= \exp\left[-\int_0^{t_k} \lambda(s) \, ds \right] \lambda(t_k). \tag{15}$$

The key point of all this is that by specifying the hazard rate, as in (8), one can derive the survivor function and the density function for the duration t_k. The survivor function accounts for right-censored observations, those that did not experience a transition. The density function accounts for observations that did experience a transition.

How this framework can be used to account for time-dependent covariates is the subject of the next section. Before proceeding to this task, I consider some specifications of the hazard rate.

The perhaps most famous specification of (8) is the exponential model

$$\lambda(t_j) = \exp(\alpha), \tag{16}$$

where the rate at duration t_j is independent of t_j. Exponentiation of α is done to ensure nonnegativity of the rate.

Several specifications allow the rate to depend on duration t_j. A simple but general specification would be

$$\lambda(t_j) = \exp(\alpha + \gamma_1 t_j + \gamma_2 \ln t_j), \tag{17}$$

in which case the exponential model obtains as a special case when $\gamma_1 = \gamma_2 = 0$; the Gompertz when $\gamma_2 = 0$; and the Weibull when $\gamma_1 = 0$ and $\gamma_2 > -1$.

In Figure 7.1 the exponential, the Gompertz, and the Weibull specifications are plotted for various parameter values. For the Weibull model the survivor function has the form

$$P[T \geq t_k] = \exp\left[-\left(\frac{1}{\gamma_2 + 1}\right) t_k^{\gamma_2+1} \exp(\alpha) \right]. \tag{18}$$

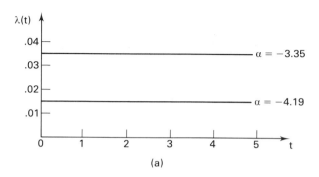

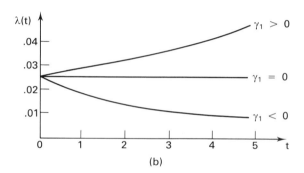

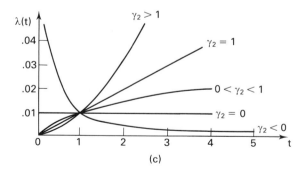

FIGURE 7.1. Shapes for the exponential, Gompertz and Weibull models: (a) exponential model, $\gamma_1 = \gamma_2 = 0$; (b) Gompertz model, $\gamma_2 = 0$, $\alpha = -3.69$; (c) Weibull model; $\gamma_1 = 0$, $\gamma_2 > -1$, $\alpha = -4.61$.

3. Dependence of the Hazard Rate on Covariates

The hazard rate at duration t_j may depend not only on t_j, as in (17), but also on explanatory variables. Let $x(t_j)$ be the vector of explanatory variables at duration t_j, where $x(t_j)$ may include lagged values of the explanatory variables.

Let the sequence of covariates from duration 0 to t_k be denoted by

$$X(t_k) \equiv \{x(s)\}_{s=0}^{s=t_k}. \tag{19}$$

The hazard rate is now defined as

$$\lambda(t_j \mid X(t_j)) \equiv \lim_{\Delta t \downarrow 0} \frac{P[t_j \le T < t_j + \Delta t \mid T \ge t_j, X(t_j)]}{\Delta t}, \tag{20}$$

giving the rate at which a transition occurs at duration t_j, given no transition before t_j, and given the covariates up until t_j. In computing (20), one conditions on the covariates up until t_j, but not to $t_j + \Delta t$. The reason is that the cause must precede the effect in time.[2]

Covariates that change over time may either be endogenous or exogenous relative to the dependent failure time process. The covariates are exogenous when they influence the probability of a failure, but are themselves not influenced by the failure time process. Otherwise they are endogenous. The relevant exogeneity condition is (see Petersen, 1983)

$$\lambda(t_j \mid X(t_k)) = \lambda(t_j \mid X(t_j)), \qquad \text{for all } k > j, \tag{21}$$

which is an extension to continuous time processes of Chamberlain's (1982) generalization of Sim's (1972) exogeneity condition for time-series data.

The condition in (21) says that future values of the covariates are not informative with respect to the probability of a present failure. If the covariates are the outcomes of the failure time process, their future values will add information about the probability of a current failure.

The condition makes sense only when the covariates are stochastic. That is, there may exist covariates whose future values will influence the likelihood of a present failure, but which are not endogenous because they are nonstochastic. An example would be an inheritance determined at birth that is to be received at the age of 20. It may influence behavior prior to age 20, but is itself not influenced by that behavior.

---

[2] Andersen and Gill (1982) provide additional technical justifications for this type of specification.

Also note that (21) does not preclude that expectations about the future may influence the probability of a transition at t_j, but expectations are to be distinguished from realizations of the future.

The survivor function, irrespective of whether the covariates are exogenous or not, given the covariates from 0 to t_k, is

$$P[T \geq t_k \mid X(t_k)] = \exp\left[-\int_0^{t_k} \lambda(s \mid X(t_k)) \, ds \right], \qquad (22)$$

that is, at each $s < t_k$, one conditions not only on the covariates up to s but also on future values of the covariates (i.e., up until t_k).

Under the assumption of exogeneity of the covariates, that is (21), the survivor function, given the sequence of covariates from 0 to t_k, becomes

$$P[T \geq t_k \mid X(t_k)] = \exp\left[-\int_0^{t_k} \lambda(s \mid X(s)) \, ds \right]. \qquad (23)$$

In the integral on the right-hand side of (23), one conditions at each $s < t_k$ only on the history of the covariates up until s.

Typically, the covariates in x change according to step functions of time. That is, the covariates stay constant at, say, $x(t_j)$, from duration t_j to t_{j+1}, at which time they jump to $x(t_{j+1})$, and so on. Suppose the covariates stay constant for k such periods of time. Suppose also that the exogeneity condition in (21) holds. In that case the survivor function in (22) or (23) reduces to (see Petersen, 1986a)

$$P[T \geq t_k \mid X(t_k)] = \exp\left[-\sum_{j=0}^{k-1} \int_{t_j}^{t_{j+1}} \lambda(s \mid X(t_j)) \, ds \right], \quad t_0 = 0, \qquad (24)$$

where each term on the right-hand side of (24) has the interpretation

$$P[T \geq t_{j+1} \mid T \geq t_j, X(t_{j+1})] = \exp\left[-\int_{t_j}^{t_{j+1}} \lambda(s \mid X(t_j)) \, ds \right], \qquad (25)$$

giving the probability of surviving beyond duration t_{j+1}, given survival at t_j, and given the covariates from 0 to t_{j+1}.

Again, it might be instructive to consider some specific examples of the hazard rate. This can easily be done within the framework of (16) and (17) of Section 2. For the exponential model, the approach is to say that the parameter α differs between groups in the sample, so individual i has parameter, say, $\alpha_i = \beta x_i(t_j)$, where $x_i(t_j)$ are the covariates at duration t_j for individual i, and β is a vector of effect parameters conforming to x_i. The rate then becomes

$$\lambda(t_j \mid x_i(t_j)) = \exp[\beta x_i(t_j)]. \qquad (26)$$

The covariates shift the rate up and down. There may be differences in the rates between individuals at a given point in time, due to differences in the covariates, and there may be intraindividual differences in the rate over time, due to intraindividual changes in x over time. If $\beta_h > 0$, the corresponding covariate increases the rate; if $\beta_h = 0$, the covariate has no effect on the rate; and if $\beta_h < 0$, the covariate lowers the rate.

In the case of the more general rate in (17), the approach is the same. Typically, one assumes that γ_1 and γ_2 do not vary between groups in the sample, but that α does, yielding the specification

$$\lambda(t_j \mid x_i(t_j)) = \exp[\beta x_i(t_j) + \gamma_1 t_j + \gamma_2 \ln t_j]. \tag{27}$$

The size of the coefficients in (26) and (27) will depend on (a) the units in which duration is measured, that is, days, weeks, and so on; (b) the units in which x is measured, as always; and (c) how often transitions occur, that is, the rate at which changes occur. Regarding (a), if duration is measured in months, the estimated rates per month will be roughly four times bigger than the estimated rates per week would have been had duration been measured in weeks, provided the weekly rates are relatively small. Except for the constant term, the coefficients in β will not be much affected by the units in which duration is measured. The reason for this is that a coefficient can be directly translated into the percentage deviation in the rate from the baseline group captured by the constant term. This percentage deviation will be unaffected by the units in which duration is measured.

To illustrate the meaning of the size of the coefficients, consider the rates in Figure 7.1(a). The figure gives the rates for two groups. The rates could have been estimated from the model $\lambda(t \mid x) = \exp(\beta_0 + \beta_1 x)$, where x is a dummy variable indicating which group the sample member belongs to. If $\beta_0 = -4.19$, as in Figure 7.1(a), it follows that $\beta_1 = .84$ $(= -3.35 + 4.19)$. The corresponding rates are $\lambda(t \mid x = 0) = \exp(-4.19) = .015$ and $\lambda(t \mid x = 1) = \exp(-3.35) = .035$. Thus, being in group 1, on the variable x, increases the rate with about 133% relative to being in group 0.

In the example just given, let $\lambda(t \mid x)$ be the rate at which any type of transition occurs. In that case, if duration is measured in months, the preceding numbers imply that the average time before a transition occurs is $E[T \mid x = 0] = 1/.01567 = 67$ months if one is in group 0, and $E[T \mid x = 1] = 1/.035 = 29$ months if one is in group 1 [see equation (35)]. On the average, members of group 0 wait more than twice as long as members of group 1 before a transition occurs. Similarly, if we look at the survivor function, the probabilities of no transition occurring before 10 months are these. For group 0, it is .859, using the standard expression for the survivor function in (13), or .858, using the approximation $\Delta t = 1$ in (11). For group 1, the probability is .704, using (13), or .700, using the approximation $\Delta t = 1$ in (11). As we see, the approximation $\Delta t = 1$ in (11) is accurate.

It is difficult to say much in general about the number of observations needed in order to estimate the parameters of a rate. The estimation procedures are usually nonlinear. My experience is that hazard rate models typically can be estimated from the same number of observations as binary logit and probit models can. Furthermore, the parameter estimates are usually stable with respect to where one starts the iteration routine. For the model in (26), the final estimates are indeed independent of the initial guesses, since the model contains no local maxima of the likelihood.

4. Comparison to Standard Regression Models

Sometimes it is remarked that the dependent variable in hazard rate models is an unobservable quantity, the hazard rate (see, e.g., Carroll, 1983, p. 430). Such remarks are incorrect. The dependent variable in hazard rate models is not the hazard rate but one of the two following, depending on one's point of view.

According to the first view, which I will call the *event history formulation,* the dependent variable is whether or not an event takes place in a small time interval t to $t + \Delta t$ (now dropping the subscripts to periods of time used in Sections 2–3). That is, it is a zero-one variable that takes the value of 1 if an event takes place in the small time interval and zero if not. We need as many such zero-one variables as there are observed time intervals.

According to the second view, which I will call the *duration formulation,* the dependent variable is the amount of time that elapses before an event or censoring occurs.

Both ways of viewing the dependent variable are equally valid, and they amount to the same thing in terms of specification, estimation, and interpretation of the models. I now explore both viewpoints.

Let

$$D(t + \Delta t) = \begin{cases} 1 & \text{if a transition occurs in the time interval } t \text{ to } t + \Delta t, \\ 0 & \text{if no transition occurs in the time interval } t \text{ to } t + \Delta t. \end{cases}$$

(28)

Using (28) we then get

$$P[D(t + \Delta t) = 1 \mid T \geq t, x] = \lambda(t \mid x)\Delta t, \qquad \text{for small } \Delta t. \qquad (29)$$

If at most one change in D can occur between t and $t + \Delta t$, it follows that

$$D(t + \Delta t) = \lambda(t \mid x)\Delta t + \varepsilon(t), \qquad \text{when } T \geq t \text{ and } \Delta t \text{ is small}, \qquad (30)$$

where $\varepsilon(t)$ is a stochastic error term with expectation 0, conditional on $T \geq t$, and x. Further, from (30), we get that

$$E[D(t + \Delta t) \mid T \geq t, x] = \lambda(t \mid x)\Delta t, \qquad \text{for small } \Delta t, \qquad (31)$$

where E is the expectations operator.

The central point here is that in (30), which captures the event history formulation, the dependent variable is whether an event takes place between t and $t + \Delta t$, given no event prior to t. There will be as many such zero-one variables as there are time units for which an individual is observed. This dependent variable takes the value of zero in all time units in which no event takes place. Only in the last time unit may it take the value of 1, if the observation is noncensored. Since we let Δt go to zero, there will be infinitely many such zero-one variables that will account for the entire duration in a state. *We can conclude that the dependent variable in event history analysis is observable.*

Once the hazard rate has been specified, the survivor function $P(T \geq t \mid x)$ and the probability density function $f(t \mid x)$ follow, by (13) and (15). The mean value of the duration T can be derived from the probability density function:

$$E(T \mid x) = \int_0^\infty sf(s \mid x)\, ds, \qquad (32)$$

from which follows that

$$T = E(T \mid x) + \varepsilon$$
$$= \int_0^\infty sf(s \mid x)\, ds + \varepsilon, \qquad (33)$$

where ε is a stochastic error term with mean zero, conditional on x.

For example, if the rate is

$$\lambda(t \mid x) = \exp(\beta x), \qquad (34)$$

we get

$$E(T \mid x) = \exp(-\beta x), \qquad (35)$$

and hence

$$T = \exp(-\beta x) + \varepsilon, \tag{36}$$

where we can estimate β by nonlinear least squares (NLLS). However, since we have specified a hazard rate, it is preferable to compute the maximum likelihood (ML) rather than the NLLS estimates.

Again, *the central point is that the dependent variable is not some unobservable instantaneous rate.* In the representation in (33), which corresponds to the duration formulation, the dependent variable is the amount of time that elapses before an event or censoring occurs. We focus on one aspect of this amount of time, the hazard rate, and we try to estimate the parameters of this rate.

5. Repeated Events

I consider the case of job mobility. Each person in the sample has held at least one job and some have held two or more jobs. The focus of the analysis will still be on the determinants of the amount of time spent in each job. A straightforward extension of the framework developed in the earlier sections will accomplish this.

Consider a person who when last observed had held m jobs with durations $t_1, t_2, \ldots, t_m$, where the last duration may be censored. Note that t_j now refers to the amount of time spent in job j, not to the duration at which period j within a job was entered, as in Sections 2–3. Let $C_m = 0$ if the last job was censored, and $C_m = 1$ if not.

Within the ML framework we need to derive the probability density of the entire job history of the person, which now is the unit of the analysis and which may consist of more than one job. Define

$$H_{j-1} \equiv \{t_g\}_{g=1}^{j-1} \qquad \text{for } j \geq 2, \tag{37}$$

which gives the sequence of durations from job 1 through job $j-1$.

The probability density of the entire job history can now be written

$$f(t_1, \ldots, t_m) = f(t_1) \prod_{j=2}^{m-1} f(t_j \mid H_{j-1})[\lambda(t_m \mid H_{m-1})]^{C_m} P[T \geq t_m \mid H_{m-1}], \tag{38}$$

where $f(t_j \mid H_{j-1})$ gives the density of the duration in job j, given the sequence of previous jobs 1 through $j-1$. The specification allows for full dependence of the duration in, say, job j on the previous job history.

decompose the destination-specific rate of transition as follows:

$$\lambda_z(t \mid z_j) = \lambda(t \mid z_j)P[Z = z \mid T = t, z_j], \tag{42}$$

that is, into the overall rate of transition times the probability of the destination state, given that a transition occurred.

The survivor function follows as

$$P[T \geq t \mid z_j] = \exp\left[-\int_0^t \sum_{z=1}^{z'} \lambda_z(s \mid z_j)\, ds\right], \tag{43}$$

which obtains by inserting the overall rate of transition in (41) into the general expression for the survivor function in equation (13).

In analyzing discrete state-space processes one can therefore either specify the destination-specific rate of transition directly, as in (40), or the overall rate of transition and the probability of the destination state, given a transition, as on the right-hand side of (42). In the first case one estimates the destination-specific rates directly. In the second case, one estimates first the overall rate of transition, using a hazard rate routine, and then the probabilities of the destination states, given a transition, using, for example, a multinomial logit model.

If one focuses on the destination-specific rate as in (40), one can, for purposes of estimation, use a hazard rate routine for estimating single-state-space processes. The estimates of each of the destination-specific rates can be obtained by separate analyses. In estimating, say, $\lambda_1(t \mid z_j)$, each transition that occurred for reason 1 is treated as noncensored, all the other observations, that is, transitions to other states or censored observations, are treated as censored. In order to estimate all the z' different rates, one just performs z' separate estimations, one for each state. This procedure is valid provided there are no restrictions on the parameters across the destination-specific rates and no unobserved variables common to, or correlated across, the rates. Each of the destination-specific rates may be given a separate functional form (Weibull, Gompertz, etc.) and may depend on different explanatory variables.

6.3. Continuous State-Space

If the state space is continuous, the framework of (40) must be modified correspondingly. Let Y be the random, and now continuous, variable denoting the state-space and let y denote a specific realization of Y. In specifying the destination-specific rate of transition, one focuses on the probability density of y being entered in a small time interval, given what has happened up to the start of the interval. The destination-specific rate of transition, $\lambda(t, y \mid y_j)$, where y_j is the state occupied immediately prior to t, is defined as (see Petersen, 1988, p. 144).

$$\lambda(t, y \mid y_j) \equiv \lim_{\substack{\Delta t \downarrow 0 \\ \Delta y \downarrow 0}} \frac{P[t \leq T < t + \Delta t, \, y \leq Y < y + \Delta y \mid T \geq t, \, y_j]}{\Delta t \, \Delta y}, \quad (44)$$

where $\lambda(t, y \mid y_j)$ is equal to zero for $y = y_j$.

The overall rate of transition follows, in a manner analogous to the discrete state-space framework in (41), by integrating over all the destination-specific rates, namely

$$\lambda(t \mid y_j) = \int \lambda(t, y \mid y_j) \, dy. \quad (45)$$

Define the density of the destination state, given a transition at duration t and given that state y_j was occupied prior to t, as

$$g(y \mid T = t, y_j) = \lim_{\substack{\Delta t \downarrow 0 \\ \Delta y \downarrow 0}} \frac{P[y \leq Y < y + \Delta y \mid t \leq T < t + \Delta t, \, y_j]}{\Delta y}. \quad (46)$$

Analogously to (42) in the discrete state-space framework, the destination-specific rate of transition can be decomposed into the overall rate of transition times the probability density of the destination state:

$$\lambda(t, y \mid y_j) = \lambda(t \mid y_j)g[y \mid T = t, y_j]. \quad (47)$$

In terms of estimation one can either focus on the destination-specific rate directly, as in (44), or on its decomposition into the overall rate of transition times the probability density of the destination state, given a transition (see Petersen, 1988, 1990), as on the right-hand side of (47).

The survivor function follows in complete analogy to (43) in the discrete state space case as (see Petersen 1990)

$$P[T \geq t \mid y_j] = \exp\left[- \int_o^t \int \lambda(s, y \mid y_j) \, dy \, ds \right]. \quad (48)$$

7. Unobserved Heterogeneity

Unobserved heterogeneity refers to the situation where some of the covariates on which the hazard rate depends are not observed. The issue is distinct from the existence of an error term in hazard rate models. As discussed in Section 4, the dependent variable in analysis of event histories can always be expressed as a sum of its mean and an error term.

To fix ideas, consider the hazard rate

$$\lambda(t \mid x, \mu) = \exp[\beta x + \delta \mu], \tag{49}$$

where μ is the unobserved variable and δ its effect; μ may for example capture whether a person is a mover or a stayer (see Spilerman, 1972).

Continuing the example of equation (30), the value of the dependent 0-1 variable $D(t + \Delta t)$ corresponding to (49) can be written as

$$D(t + \Delta t) = \exp[\beta x + \delta \mu]\Delta t + \varepsilon(t), \qquad \text{when } T \geq t \text{ and } \Delta t \text{ is small}, \quad (50)$$

where $\varepsilon(t)$ is an error term with mean of zero, conditional on x, μ, and $T \geq t$. Equation (50) shows that μ is distinct from an error term $\varepsilon(t)$.

The problem that the unobserved variable μ gives rise to is that even if its mean, conditional on x, is zero, that is, $E(\mu \mid x) = 0$, the mean of $\exp(\delta \mu)$, conditional on x and t, will typically be different from one (by Jensen's inequality; see, e.g., Degroot, 1970, p. 97). For example, if μ is gamma distributed with mean 1 and variance $1/\sigma$, the mean value of the rate, unconditional on μ, for those who had not experienced an event by duration t becomes (Tuma and Hannan, 1984, p. 179)

$$\lambda(t \mid x) = [\sigma \exp(\beta x)]/[t \exp(\beta x) + \sigma], \tag{51}$$

which depends on t. The rate in (50) is independent of t, but when μ is not observed, the rate, unconditional on μ, ends up depending on t, and the duration dependence is negative.

Therefore, when an unobserved variable enters into the hazard rate, as in (49), it typically does not cancel out when it comes to estimation, as the error term ε does. In order to avoid biases in the estimates, one must take account of the unobservable in the estimation procedure.

Several solutions to the problem of unobserved heterogeneity have been proposed, but two are prominent in the literature (see Yamaguchi, 1986).

The first solution assumes, as before, that μ is a random variable, commonly referred to as the *random-effects procedure*. The main approach in that case is to impose a specific distribution on μ, say, normal, lognormal, or gamma. In terms of estimation, one then first derives the likelihood for the observed history of durations on an individual, conditional on observed and unobserved explanatory variables. Thereafter one uses the imposed distribution of the unobservable to compute the mean of the likelihood when the unobserved variable is not taken into account. That is, one computes the average value of the likelihood, where the averaging is done over all possible values of the unobserved variable. This average depends on the observed variables, on the parameters of the hazard rate, and on the parameters of the distribution of the unobservable. This procedure is repeated for each individual in the sample. Thereafter one maximizes the likelihood of the sample by standard procedures (see Lancaster, 1979).[3]

An example of this approach obtains if we assume μ in (49) to be binomially distributed with parameter p, which gives the probability of a person being, for example, a mover or a stayer. The likelihood of an observation, after having "expected out" the unobservable, then is

$$L^* = p[\exp(\beta x + \delta)]^C \exp[-t \exp(\beta x + \delta)]$$
$$+ (1 - p)[\exp(\beta x)]^C \exp[-t \exp(\beta x)], \quad (52)$$

where C, as before, is a censoring indicator.

Neither theory nor data gives much guidance when choosing the distribution for μ. Heckman and Singer (1984a) therefore developed an estimator of the hazard rate that is nonparametric with respect to the distribution of μ. There is not much experience with this estimator (see, however, Trussel and Richards, 1985).

The second main solution is to treat μ as a fixed variable (Chamberlain, 1985), referred to as the *fixed-effects procedure*. Although appealing in that few assumptions need to be imposed on μ, this procedure has the drawback that it applies only to processes where the event is repeated over time and where one has observed at least two transitions on some of the individuals in the sample. Furthermore, and perhaps more restrictive, only the effects of covariates that change over time can be estimated.

8. Theoretical Models Generating Event Histories

Often the researcher starts with a theoretical model for how the data were generated. The objective of the empirical analysis is then to use the data and the statistical techniques to estimate the parameters of the underlying theoretical model. In this section I consider a simple example of such a situation, drawn from sociological research on intragenerational mobility (see, e.g., Sørensen, 1979).

---

[3] A variant of this procedure is the so-called EM algorithm (see Dempster and Laird, 1977).

A person with socioeconomic status (or earnings) y_j receives job offers at a rate $\lambda(t \mid y_j)$, where t is the time elapsed since y_j was entered. The offer y comes from a distribution with density $g(y \mid t, y_j)$.

Assume that the worker maximizes socioeconomic status and that there are no costs of changing jobs (an assumption that is straightforward to relax). Assume also that all job changes are voluntary. If so, a person will accept any offer for which $y > y_j$; otherwise he or she rejects it. Let $D = 1$ denote that the offer is acceptable, that is, $y > y_j$, and $D = 0$ otherwise.

The rate $\lambda(t \mid y_j)$ at which an upward change in socioeconomic status occurs then equals

$$\lambda_1(t \mid y_j) = \lambda(t \mid y_j)P[D = 1 \mid T = t, y_j]. \tag{53}$$

That is, the rate of an upward shift equals the overall rate at which offers arrive times the probability that the offer is acceptable, given the arrival of an offer.

The researcher might be interested in recovering (i.e., estimating) both the overall rate and the distribution of the offers. A specific example in which the researcher can achieve both objectives may help clarify the ideas. Let the arrival rate of offers be given by

$$\lambda(t \mid y_j) = \exp[\beta_0 + \beta_1 y_j], \tag{54}$$

and let the density of an offer y, given an arrival at duration t, be

$$g(y \mid t, y_j) = \xi \exp[-\xi y]. \tag{55}$$

The rate at which an upward shift occurs then equals

$$\lambda_1(t \mid y_j) = \exp[\beta_0 + \beta_1 y_j] \exp[-\xi y_j]$$
$$= \exp[\beta_0 + (\beta_1 - \xi)y_j], \tag{56}$$

and the density of an accepted offer, y_{j+1}, at duration t_{j+1} is

$$g(y_{j+1} \mid T = t_{j+1}, y_j, y_{j+1} > y_j) = \xi \exp[-\xi(y_{j+1} - y_j)]. \tag{57}$$

From the data on the durations before an upward shift, it is clear that we can identify β_0 and a reduced form of β_1, namely, $\psi \equiv \beta_1 - \xi$, but not the structural coefficient β_1, as is seen from (56).

However, from the data on the values of socioeconomic status before and after an upward shift, we can identify ξ, as is seen from (57). And then, using the estimates of ξ and of the reduced form parameter ψ, we can identify the structural parameter β_1, from $\beta_1 \equiv \psi + \xi$.

The example shows that it may be useful to consider some underlying process that generates the data. The estimates obtained from the data will have interpretations relative to the underlying model. Specifically, in the example given, focusing only on the rate of upward shifts, that is, equation (56), will not enable one to recover the parameters of interest. The reduced form estimates are mongrels of the parameters of the rate at which offers arrive and the parameters of the distribution of the offers.

9. Empirical Analysis

I present an example of a two-state hazard rate model. The data are taken from the personnel records of a large U.S. insurance company. For each employee in the company we know the date he or she entered the company and the dates of all movements within the company up until the end of the study, or the date the person quit the company (end of study is December 1978). I present estimates of the rates of promotion and of departure from the company.

The company is hierarchically organized into salary grade levels, from grade 1 (the lowest) to grade 20 (the highest). On top of those ranks are the various ranks of vice-presidents. I analyze the amount of time that elapses from the date an employee enters into a given salary grade level until a promotion, departure, or censoring occurs. Demotions are very rare. Hence, the two-state model, distinguishing promotions and departures, in practice exhausts the possibilities. For further details on the data see Petersen and Spilerman (1989). I restrict the analysis to lower-level clerical employees in the company, all of whom are employed in salary grade levels 1–6.

Let t denote the time that has elapsed since a salary grade level was entered, and let $x(t)$ be the covariates evaluated at duration t, where $x(t)$ includes the constant 1.

The rate of leaving the company is specified as a Weibull model

$$\lambda_d(t) = \exp[\beta_d x(t) + \gamma_d \ln t], \tag{58}$$

where $\gamma_d > -1$ and β_d is a parameter vector conforming to $x(t)$. When $\gamma_d > 0$, the rate increases with t; when $\gamma_d = 0$, the rate is independent of t; and when $\gamma_d < 0$, the rate declines with t [see Figure 7.1(c)].

The rate for getting promoted is specified as

$$\lambda_p(t \mid x(t)) = \frac{(\gamma_p + 1)t^{\gamma_p} \exp[\beta_p x(t)]}{1 + t^{\gamma_p + 1} \exp[\beta_p x(t)]}, \tag{59}$$

where $\gamma_p > -1$ and β_p is a vector of parameters conforming to $x(t)$. This is a nonproportional hazards specification of the so-called log-logistic model. When $\gamma_d \leq 0$, the rate declines monotonically with duration t. When $\gamma > 0$, the rate first increases with duration in grade, then reaches a peak, and thereafter declines. That is, the rate is a bell-shaped function of time.

The variables in the x-vector are sex, race (white, black, Hispanic, or

TABLE 7.1. Estimates of the Parameters of the Multistate Hazard Rate Model in Equations (58) and (59) (estimated standard errors in parentheses)

	Destination state	
	Departure from company (Equation 58)	Promotion within company (Equation 59)
Constant	−2.064 (.042)	−6.056 (.074)
Duration (in grade)	−.149 (.006)	1.044 (.012)
Race[a]		
Black	.003 (.019)	−.197 (.028)
Asian	.078 (.057)	.231 (.091)
Spanish	−.141 (.031)	−.110 (.043)
Female (= 1)	−.294 (.026)	−.357 (.037)
High school or more (= 1)[b]	−.177 (.030)	.848 (.050)
Location (1 = home)	−.587 (.018)	.261 (.024)
Salary grade level[c]		
2	−.360 (.021)	−.422 (.037)
3	−.589 (.021)	−.743 (.364)
4	−.907 (.024)	−1.091 (.038)
5	−1.335 (.032)	−1.318 (.041)
6	−1.403 (.056)	−1.696 (.063)
− Log-likelihood[d]	77986.2	84232.4

Note. The parameters of the multi-state hazard in equations (58) and (59) were estimated by the Method of Maximum likelihood, using the algorithm described in Petersen (1986b). The number of employees contributing to the estimation is 25,671. The number of departures is 17,490 and the number of promotions is 19,449. See Section 9 for description of the data (see also Petersen and Spilerman, 1989).
[a] Excluded group: White and Native Americans.
[b] Excluded group: Less than high school education.
[c] Excluded group: salary grade level 1.
[d] The log-likelihood is partioned into two parts, one for each of the two rates.

Asian), educational level (less than high school, or high school or more), the salary grade level currently occupied (entered as five dummy variables), and the location of the company in which the employee works (home office or elswhere). The location variable is treated as time-dependent within as well as between salary grade levels.

The estimates of the parameters are obtained by the method of maximum likelihood, using the algorithm described in Petersen (1986b, Appendix A). A full listing and explanation of the algorithm can be found in Blossfeld, Hamerle, and Mayer (1989, Chap. 6). The parameter estimates are given in Table 7.1. I restrict the discussion to the four most striking results in Table 7.1.

First, we see that both the rates of departure and of getting promoted decline strongly with the salary grade level occupied. The higher up in the company, the less likely an employee is to leave and the longer it takes to get promoted. This probably means that the benefits accruing from being in the upper echelons of the salary grade levels for lower-level clerical employees must outweigh the drawback of the lower promotion rates once these grades have been reached. Otherwise, one would expect departure rates to increase with salary grade level.

Second, duration in a salary grade level has a negative effect on the rate of leaving the company, and it has a bell-shaped effect on the rate of promotion. In the initial months in a salary grade level, the rate of promotion increases; it then reaches a peak, whereafter it declines.

Third, the rate of promotion is higher in the home office than elsewhere, whereas the rate of departure is lower. When opportunities for advancement are high, quit rates are lower, given the level of already obtained achievement (i.e., the salary grade level).

Finally, the race, sex, and educational effects on the promotion rate are as one would expect. Women and blacks have lower promotion rates than the other groups. Having a high school degree or more education increases the rate of promotion.

10.   Concluding Remarks

I have presented an overview of some central themes in analyzing event histories. Some topics have been omitted, four of which, in my opinion, are important. I conclude with some brief remarks on these topics.

First, I have not discussed likelihood construction in the presence of both left- and right-censored data. The data are left-censored when we know that a person was in a state as of, say, τ_0, but we do not know the date prior to τ_0 at which the state actually was entered. Left censoring is considerably more complicated to deal with than right-censoring, except for the simple case of the exponential model of equation (16). Tuma and Hannan (1984, pp. 128–135) discuss left-censoring.

Second, I have not discussed problems that arise in connection with imprecise measurement of the recorded dates of events. We may know that an event took place between t_j and t_{j+1}, but not the exact time within that interval. Often, researchers then assume that the event took place at $t^* = t_j + (t_{j+1} - t_j)/2$, treating t^* as the exact date of the event. This will give rise to some time aggregation bias in the estimates, and one can do better by constructing an estimator that takes into account the grouped nature of the data (see Thompson, 1977; Allison, 1982; Petersen, 1991).

Third, I have not discussed a host of problems that arise in connection with specific sampling plans for collecting event histories. Suppose one samples only the last two events on each individual in a target sample. In principle, one should then construct an estimator that adjusts for the fact that only portions of the life histories are observed. Informative, albeit difficult, discussions of these issues are found in Hoem (1985), Heckman and Singer (1984b, Sect. 8), and Ridders (1984).

Finally, I have not discussed the partial likelihood principle for estimating the parameters of so-called proportional hazards models, an example of which is (58) but not (59). This principle is powerful in that one can estimate the effects of the covariates without assuming anything about the effect of duration in the state, other than that it enters multiplicatively into the hazard (see, e.g., Cox, 1975).

Appendix A.   Proof of the Survivor Function in Equation (13)

We shall prove

$$P[T \geq t_k] = \lim_{\substack{k \to \infty \\ \Delta t \downarrow 0}} \prod_{j=1}^{k} [1 - \lambda(t_{j-1})\Delta t] = \exp\left[-\int_0^{t_k} \lambda(s)\, ds \right]. \qquad (13)$$

In order to do this we will need the well-known approximation

$$1 - \lambda(t_{j-1})\Delta t \approx \exp[-\lambda(t_{j-1})\Delta t], \qquad \text{for } \lambda(t_{j-1})\Delta t \text{ positive and small.}$$

$$(A1)$$

Insert the right-hand side of (A1) into each term on the right-hand side of the first equality in (13). This yields

$$\lim_{\substack{k\to\infty \\ \Delta t \downarrow 0}} \prod_{j=1}^{k} [1 - \lambda(t_{j-1})\Delta t] = \lim_{\substack{k\to\infty \\ \Delta t \downarrow 0}} \prod_{j=1}^{k} \exp[-\lambda(t_{j-1})\Delta t]$$

$$= \lim_{\substack{k\to\infty \\ \Delta t \downarrow 0}} \left[- \sum_{j=1}^{k} \lambda(t_{j-1})\Delta t \right]$$

$$= \exp\left[- \lim_{\substack{k\to\infty \\ \Delta t \downarrow 0}} \sum_{j=1}^{k} \lambda(t_{j-1})\Delta t \right]. \qquad (A2)$$

Next, we need the definition

$$\lim_{\substack{k\to\infty \\ \Delta t \downarrow 0}} \sum_{j=1}^{k} \lambda(t_{j-1})\Delta t \equiv \int_{0}^{t_k} \lambda(s)\, ds, \qquad \text{when } t_0 = 0 \text{ and } t_k \text{ is fixed}, \quad (A3)$$

where the left-hand side of (A3) is nothing but the definition of an integral, often called a Riemann-Stieltjes integral (see Kolmogorov and Fomin, 1970, pp. 367–368).

The expression for the survivor function in (13) follows from inserting the definition in (A3) into (A2).

Appendix B. Computer Programs for Estimating Hazard Rate Models

Several computer programs are available for estimating hazard rate models. For example, estimation routines are available in BMDP (see BMDP 1985), GLIM (see Baker and Nelder, 1978), and in the special-purpose program RATE of Nancy B. Tuma (see the description in Blossfeld, Hamerle, and Mayer, 1989, Chap. 6).

Excellent descriptions of several of the available programs can be found in Blossfeld, Hamerle, and Mayer (1989, Chaps. 5–6). They present extensive examples and comparisons between BMDP, GLIM, RATE, and the routine developed in Petersen (1986b), which was used in estimating the parameters in Table 7.1. Allison (1984, Appendix C) also discusses computer programs. I refer interested readers to these two sources.

References

Allison, P. D. (1982). "Discrete-time methods for the analysis of event histories." In S. Leinhardt (Ed.), *Sociological methodology 1982* (pp. 61–98). San Francisco: Jossey-Bass.

Allison, P. D. (1984). *Event history analysis: regression for longitudinal event data.* Beverly Hills: Sage.

Andersen, P. K., and Borgan, Ø. (1985). "Counting process models for life history data: a review (with discussion)." *Scandinavian Journal of Statistics.* **12:** 97–158.

Andersen, P. K., and Gill, R. D. (1982). "Cox regression model for counting processes: a large sample study," *Annals of Statistics.* **10:** 1100–1120.

Apostol, T. M. (1967) *Calculus,* Vol. 1, 2nd ed. New York: Wiley.

Baker, R. J., and Nelder, J. A. (1978). *The GLIM system.* Oxford, U.K.: Numerical Algorithms Group.

Blossfeld, H.-P., Hamerle, A., and Mayer, K. U. (1989). *Event history analysis.* Hillsdale, N.J.: Lawrence Erlbaum.

BMDP Statistical Software (1985). Berkeley, Calif.: University of California Press.

Carroll, G. C. (1983). "Dynamic analysis of discrete dependent variables: A didactic essay." *Quality and Quantity.* **17:** 425–460.

Chamberlain, G. (1982). "The general equivalence of Granger and Sims causality," *Econometrica.* **50:** 569–581.

Chamberlain, G. (1985). "Heterogeneity, omitted variable bias, and duration dependence." In J. J. Heckman and B. Singer (eds.), *Longitudinal analysis of labor market data* (pp. 3–38). New York: Cambridge University Press.

Cox, D. R. (1975). "Partial likelihood," *Biometrika.* **62:** 269–276.

DeGroot, M. H. (1970). *Optimal statistical decisions.* New York: McGraw-Hill.

Dempster, A. P., Laird, N. and Rubin, D. B. (1977). "Maximum likelihood from incomplete data via the EM algorithm," *Journal of the Royal Statistical Society,* Ser. B, **39:** 1–38.

Flinn, C. J., and Heckman, J. J. (1983). "Are unemployment and out of the labor force behaviorally distinct labor force states," *Journal of Labor Economics.* **1:** 28–42.

Heckman, J. J., and Singer, B. (1984a). "A method for minimizing the impact of distributional assumptions in econometric models for duration data," *Econometrica.* **52:** 271–320.

Heckman, J. J., and Singer, B. (1984b). "Econometric duration analysis," *Journal of Econometrics.* **24:** 63–132.

Hernes, G. (1972). "The process of entry into first marriage," *American Sociological Review.* **37:** 173–82.

Hoem, J. M. (1985). "Weighting, misclassification, and other issues in the analysis of survey samples of life histories." In J. J. Heckman and B. Singer (eds.), *Longitudinal analysis of labor market data* (pp. 249–283). New York: Cambridge University Press.

Kalbfleisch, J. D., and Prentice, R. L. (1980). *The statistical analysis of failure time data.* New York: Wiley.

Kolmogorov, A. N., and Fomin, S. V. (1970). *Introductory real analysis.* New York: Dover.

Lancaster, T. (1979). "Econometric methods for the duration of unemployment," *Econometrica.* **47:** 939–956.

Petersen, T. (1983). "Simultaneous equations models for analysis of event-history data." University of Wisconsin, Madison: Center for Demography and Ecology, Working Paper #83–48.

Petersen, T. (1986a). "Fitting parametric survival models with time-dependent covariates," *Journal of the Royal Statistical Society, Ser. C.* **35:** 281–288.

Petersen, T. (1986b). "Estimating fully parameteric hazard rate models with time-dependent covariates. Use of maximum likelihood," *Sociological Methods and Research.* **14:** 219–246.

Petersen, T. (1988). "Analyzing change over time in a continuous dependent variable: specification and estimation of continuous state space hazard rate models." In C. C. Clogg (ed.), *Sociological methodology 1988* (pp. 137–64). Washington, D.C.: American Sociological Association.

Petersen, T. (1990). "Analyzing continuous state space failure time processes: two further results." *Journal of Mathematical Sociology* **15** (3–4)**:** 247–257.

Petersen, T. (1991). "Time aggregation bias in continuous time hazard rate models." In P. V. Marsden (ed.), *Sociological methodology 1991* **21.** Cambridge, MA: Basil Blackwell.

Petersen, T., and Spilerman, S. (1989). "Job-quits from an internal labor market." In K. U. Mayer and N. B. Tuma (eds.), *Applications of event history analysis in life course research* (Chap. 4). Madison, Wis.: University of Wisconsin Press.

Ridders, G. (1984). "The distribution of single-spell duration data." In G. R. Neumann and N. C. Westergard-Nielsen (eds.), *Studies in labor market dynamics* (pp. 45–73). New York: Springer-Verlag.

Sims, C. A. (1972). "Money, income, and causality," *American Economic Review.* **62:** 540–52.

Sørensen, A. B. (1979). "A model and a metric for the analysis of the intragenerational status attainment process," *American Journal of Sociology.* **85:** 361–384.

Spilerman, S. (1972). "Extensions of the mover stayer model," *American Journal of Sociology.* **78:** 599–627.

Thompson, W. A., Jr. (1977). "On the treatment of grouped observations in life studies," *Biometrics.* **33:** 463–470.

Trussel, J., and Richards, T. (1985). "Correcting for unmeasured heterogeneity using the Heckman–Singer procedure." In N. B. Tuma (ed.), *Sociological methodology 1985* (pp. 242–276). San Francisco: Jossey-Bass.

Tuma, N. B. (1976). "Rewards, resources and the rate of mobility: a nonstationary multivariate stochastic model," *American Sociological Review.* **41:** 338–360.

Tuma, N. B., and Hannan, M. T. (1984). *Social dynamics: models and methods.* Orlando, Fla.: Academic Press.

Vaupel, J. W., Manton, K. G., and Stallard, E. (1979). "The impact of heterogeneity in individual frailty on the dynamics of mortality," *Demography.* **16:** 439–454.

Yamaguchi, K. (1986). "Alternative approaches to unobserved heterogeneity in the analysis of repeatable events," In N. B. Tuma (ed.), *Sociological methodology 1986* (pp. 213–249). San Francisco: Jossey-Bass.

Chapter 8

Linear and Nonlinear Curve Fitting

DAVID THISSEN

Department of Psychology
University of Kansas
Lawrence, Kansas

R. DARRELL BOCK

Departments of Education and Behavioral Science
The University of Chicago
Chicago, Illinois

Abstract

This chapter treats curve fitting in the context of the measurement and description of human growth in stature from birth to maturity. The treatment of stature is illustrative; growth curve fitting in other contexts would follow steps similar to those described here. First, on some theoretical and empirical basis, an appropriate family of curves is specified, usually with extensive exploratory data analysis. After some plausible models have been specified, further data are collected. Then the population distributions of the parameters for individual growth can be estimated by marginal maximum likelihood and incorporated in multi-level statistical procedures for fitting at the case level. Finally, if prediction is desirable, the relationship of other variables to the parameters of the growth model can be investigated by multivariate statistical methods in order to find collateral information that improves prediction. If sufficient longitudinal data are available to evaluate the results, a formal system of prediction with specified accuracy can then be developed.

289

"Applications of mathematics are always complicated by the obligation to be true to the subject matter treated as well as the mathematics." *Frederick Mosteller and John Tukey*

1. Models for Growth

In the spirit of the quote from Mosteller and Tukey (1977), we will keep our discussion true to the subject matter of this book by treating curve fitting in the context of one of the standard problems in the study of human development—the measurement and description of human growth in stature from birth to maturity. Although height, whether measured as standing height or recumbent length, is perhaps not the ideal scientific variable, it has proved invaluable as an index of developmental progress in childhood and of the health and nutritional status of populations (see Tanner, 1978). It is also occupationally relevant, especially for jobs requiring physical strength, and has always been considered for military service. Much of our data on secular changes in average height derives from military records. Perhaps most important, height is a socially significant characteristic of persons, so much so that parents will often seek medical intervention to insure that their children will grow to a mature height within the normal range.

For these many reasons, the literature contains a great number of growth studies, cross-sectional and longitudinal, invariably including height among the developmental measures (Eveleth and Tanner, 1976). A collateral literature discusses the statistical methods for summarizing and analyzing data from such studies. The main topic of the latter is the fitting of growth curves expressing height, or other measures, as a function of chronological age. Considering that even these statistical studies are too extensive to cover fully in a single chapter, we have chosen to limit our discussion to the approach to curve fitting that we believe to be most useful in this context, namely, the estimation of parametric, structural models for growth as a function of age.[1] Although nonstructural methods of representing growth, such as "Tuckerizing," considered in Chapter 11 of Volume II, and spline function or kernel estimation are generally satisfactory if the data are regular (observed at

---

[1] Here we use the term *structural* to refer to models representing growth as a prespecified mathematical function of age. Nonstructural (alternative) methods are frequently parametric, but their mathematical form is not prespecified as $y = f(t)$, where the only unknowns in $f(t)$ are the parameters.

uniformly fixed time intervals) and complete (see Gasser et al., 1985), structural modeling by the methods discussed here is better adapted to the incomplete and irregular data that is typical of most growth studies. If the model is well chosen, the structural approach also has a potential for biological interpretation that is largely lacking in the other more purely descriptive methods. Moreover, as we discuss in Section 3, it supports prediction of growth in a way that the other approaches do not.

Before modern computing equipment became available, there were compelling reasons for restricting parametric models for growth to those that are linear in their coefficients. Under normal error assumptions, the parameters of such models have simple sufficient statistics that may be computed in one pass through the data. Unfortunately, this is not the case for most nonlinear models, and the computational labor of efficient estimation has limited their use until relatively recently. Even now their statistical treatment is not very well known to applied workers, many of whom continue to rely on general systems of linear modeling such as finite Fourier series or polynomials in the powers of the age variable. But long-term developmental data, especially those involving growth to an asymptote at maturity, are typically not well suited to these types of representation, for they tend to produce oscillations where none could plausibly exist. For these reasons, and also because the procedures for fitting the nonlinear models apply as well to linear models (and converge in one iteration when so applied), we will further limit our discussion to estimation for *nonlinear* growth models.

Our goal is to apply nonlinear models to the fitting of curves to longitudinal growth data from individual cases. We can then use the individual's fitted curve both for smoothing the growth record by plotting the function and for prediction by extrapolating the function beyond the range of the data. In addition, if we wish to compare the growth patterns of two or more groups (e.g., boys and girls), we may take the parameter estimates for each person as a description of his or her entire growth pattern, and subject those parameter estimates to multivariate analysis of variance. This comparison of descriptive parameter estimates is much better justified than the seemingly simpler alternative of averaging the measurements at each time point. Not only does averaging the data require the measurements to be made at exactly the same time points for all cases, it also obscures the functional form of the individual growth curves. Merrell (1931) showed that, if individual growth follows a particular nonlinear growth curve, the average values of the response variable as a function of age will in general follow a differently shaped curve. Thus it is important that when we compare groups we analyze the mean curve, not the curve for the mean of the data. Averaging the parameter estimates and plotting the resulting function provides a typical curve with the correct shape.

Taking the parameter estimates of the individual cases as data for a conventional group-level multivariate analysis assumes, however, that the case-level data are sufficiently extensive that the error of parameter estimation is small relative to the between-case variation. Under this condition, the estimates may reasonably be used as sufficient statistics for the population parameters. This is a favorable situation for data analysis because exact, small-sample (in the number of cases) statistical theory then applies in the group-level analysis.

If the data for each individual case are poor, however, a more efficient analysis at the group level can be carried out by the *marginal maximum likelihood* (MML) method, in which group-level effects are estimated directly from the original data. This method requires, however, that the the number of cases be large. In effect, by adopting the MML method we can compensate for unreliable or incomplete measurement at the case level by increasing the number of cases sampled. The larger sample of cases makes up at the population level for randomly unreliable or incomplete measurement at the case level. It cannot, of course, correct systematically biased measurement; only more careful attention to the measurement technique can do that.

As we show in Section 2, the MML methods can improve the estimation of population means and covariance matrices for the model parameters by including, in the process of fitting the individual curves, information about the distribution of the parameters in the population. These techniques, which are becoming widespread in applied statistics, have been variously called *James-Stein estimation* (after James and Stein, 1961), *Bayesian hyperparameter models* (Lindley and Smith, 1972), and *empirical Bayes techniques* (Maritz, 1970). By whatever name, instead of directly "averaging" parameter estimates, these methods use Bayes rule to combine the information from the cases in the sample with prior information about the population to improve both the curve fitting at the case level and the estimation of means and covariances at the population level.

A more descriptive name for the procedure, described in Section 2, is *multilevel analysis* or, in this instance, *two-level analysis*. As the latter name implies, there are two levels of sampling involved—occasions within individual cases, and cases within the population. Information from both levels contributes to the estimation of parameters at both levels (see Dempster, Rubin, and Tsutakawa, 1981). The fact that Bayes' rule is involved in combining the information is less important than that there are multiple levels. Multilevel analysis can be done without using Bayes rule; mixed model analysis of repeated measures is, for example, a form of two-level analysis.

In Section 3, we describe how multilevel analysis can be used to move from growth curve *fitting* to growth curve *prediction*. If we know, or have estimated, the population distribution of the model parameters, we can make an estimate of the growth curve for an individual case even when there is very little data for that case. Such estimation amounts to a prediction, or "attribution," of data that have not been observed. The prediction is a compromise between the small amount of data for the case and the average curve for the population from which the case is sampled. It employs Bayes rule to make the compromise optimally in the sense of minimizing the mean square errors of prediction averaged over the population.

1.1. *Some Families of Growth Curves*

Work on fitting growth curves to human stature is almost as old as the present century. Robertson (1908) appears to have been the first to suggest that the logistic curve could be fitted to growth. In his early work, he borrowed the logistic curve from chemical kinetics, where it is used to represent the percent of completion of a self-limiting autocatalytic reaction. Robertson apparently took seriously the idea that the curve applied to biological growth because growth is chemically based. In particular, it would be assumed to describe the competing processes of cell division and ossification at the growth epiphysis, as the latter increasingly limits the number of cells available for division.

C. P. Winsor (1932) suggested the Gompertz function as an alternative to the logistic curve, both of which are solutions to simple differential equations describing the rate of growth. Preece and Baines (1978) proposed solutions of somewhat more complex differential equations as a family of "modified logistic" models for human growth in stature.

Jenns and Bayley (1937) suggested a curve, named after them, for early childhood growth, and Count (1943) offered an alternative in which the complete human growth curve was represented as a combination of three different curves. Earlier, Burt (1937) had employed a sum of three logistics to fit a curve for the average growth of a group of children. The

TABLE 8.1.   Some Growth Curves for Stature

Early childhood to maturity

Triple-logistic

  Bock and Thissen, 1976
  Bock and Thissen, 1980 $\hat{y} = \dfrac{a_1}{1 + e^{-b_1(t - c_1)}} + \dfrac{a_2}{1 + e^{-b_2(t - c_2)}} + \dfrac{a_3}{1 + e^{-b_3(t - c_3)}}$
  Bock, 1986
  Thissen and Sykes, 1984

Double-logistic

  Bock et al., 1973
  Rarick et al., 1975 $\hat{y} = \dfrac{a_1}{1 + e^{-b_1(t - c_1)}} + \dfrac{a_2}{1 + e^{-b_2(t - c_2)}}$
  Johnston et al., 1976
  Thissen et al., 1976

Model 1 $\hat{y} = y_1 - \dfrac{2(y_1 - y_c)}{e^{b_1(t - c)} + e^{b_2(t - c)}}$
  Preece and Baines, 1978

Model 3 $\hat{y} = y_1 - \dfrac{4(y_1 - y_c)}{e^{b_1(t - c)} + e^{b_2(t - c)} + e^{b_3(t - c)}}$
  Preece and Baines, 1978

Adolescence

Constant + Gompertz

  Marubini et al., 1971 $\hat{y} = a_1 + a_2 e^{-e^{-b(t - c)}}$
  Marubini et al., 1972

Constant + logistic

  Marubini et al., 1971 $\hat{y} = a_1 + \dfrac{a_2}{1 + e^{-b(t - c)}}$
  Marubini et al., 1972

Infancy to early childhood

Jenns Model

  Jenns and Bayley, 1937
  Manwani and Agarwal, 1973 $\hat{y} = a + bt - e^{c + dt}$
  Berkey, 1982a
  Berkey, 1982b

Count Model

  Count, 1943 $\hat{y} = a + bt + c \log t$
  Berkey, 1982a

---

triple-logistic function we introduced in 1976 is a simpler version of that model (Bock and Thissen, 1976, 1980).

Table 8.1 summarizes some of the parametric functions that have been proposed as models for human growth in stature. The tabulation is not exhaustive, as there have been literally dozens of proposals (see Goldstein, 1979, Chap. 4). In each of these models, $\hat{y}$ is the fitted value of stature at age t. The variables denoted by a, b, c, or d (subscripted in the multicomponent curves) are the parameters, and subscripted values of y represent the value of stature at particular points in growth.

Although the parameters of each of the curves in Table 8.1 have some more or less plausible interpretation as descriptions of the growth process, we will present here only our interpretation of the parameters of the Bock and Thissen (1976, 1980) triple-logistic model. The model assumes that human growth from about one year of age to maturity is the sum of three self-limiting components referred to as the "infantile," "middle-childhood," and "adolescent" components, respectively. Each of the components k ($k = 1, 2, 3$) is characterized by three parameters:

a_k, the component's contribution to mature stature

b_k, a parameter proportional to the maximum growth velocity of the component (maximum rate of growth is $ab/4$ cm/yr)

c_k, the age in years at which the maximum growth rate occurs

Figure 8.1 shows the components and triple logistic curves for a hypothetical "average" boy and girl from the data of the Berkeley Guidance Study (Tuddenham and Snyder, 1954). The second, or mid-childhood, component contributes most to mature height and therefore most reflects the familial component of overall linear size. Reaching its maximum around seven years, corresponding to the maturing of the adrenals ("adrenarchy"), this component models the mid-childhood growth spurt that can be seen, especially in boys, if not obscured by early pubertal growth.

The shape of the growth curve from one to six years is modified from the logistic form of the second component by the contribution of the first, or infantile, component. The latter accounts for changes in growth channels during the first three years of life associated with variation in gestational age at birth and corresponding catch-up growth (Dunn, Hughes, and Schulzer, 1983). This component reaches its asymptote by about six years in those children who exhibit the mid-childhood maximum. The transition from early to middle growth can be located by the corresponding zero of the second derivative (acceleration) of the triple-

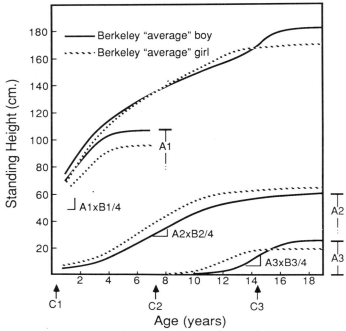

FIGURE 8.1.   The triple-logistic components and growth curves constructed from the average growth parameters for the Berkeley Guidance Study sample. (Reprinted with permission from Taylor and Francis, Bock, R. D. "Unusual growth patterns in the Fels data" in A. Demirjian, ed., *Human growth: multidisciplinary review,* Figure 1 (1986).)

logistic model. (The TRIFIT program finds these zeros; Bock and Thissen, 1986). It is because of individual differences in growth during the first six years of life (Reed and Stuart, 1959), and the occurrence of the mid-childhood growth velocity maximum in about 50% of cases, that the triple-logistic model fits much better than the double-logistic, which was an early attempt (Bock, et al., 1973) to model the complete growth curve (Bock and Thissen, 1976; el Lozy, 1978).

As the gonads mature at puberty, growth of cartilage is stimulated and the adolescent growth spurt begins, only to be limited by the competing process of epiphyseal closure (Preece, 1986). Closure begins distally, so hands and feet mature before arms and legs, which mature before trunk. Estrogen accelerates closure of the epiphyses more than testosterone, which accounts for women having generally greater trunk length relative to leg length than do men. Pediatricians can take advantage of this effect by administering estrogen in early adolescence to decrease mature stature. In contrast, castration of males, as practiced in the 17th and 18th-century Italian opera, resulted in unusually tall and long-stemmed men (see Peschel and Peschel, 1987), apparently because adrenal androgens are sufficient to stimulate growth of cartilage but not to effect closure. In those cases, some other process terminated growth without epiphyseal fusion (Tanner, 1978).

Depending on nutritional status (see Frisch, 1988), adolescent growth in girls reaches its maximum around 12 years and in boys around 14 years. Almost all girls reach mature height by 18 and boys by 20 years. Height then remains essentially constant until after 40 or 50, when a slow decrease begins due to shrinking of the intervertebral disks and greater curvature of the spine. The latter effects are greater in women than in men (Chandler and Bock, 1988).

Studies of the relationship between epiphyseal closure, as seen in hand-wrist radiograms, and growth in stature show that the timing of puberty has an important effect on mature height (Bock, 1986). The contribution of the adolescent component itself does not vary so greatly from case to case, but if it comes early, the contribution of the mid-childhood component is cut short, and a smaller mature height with reduced stem length relative to trunk results. If it is late, the contribution of the mid-childhood component continues longer than normal, and a more eunuchoid body form tends to be seen.

The distribution of the mean squared residuals of observed growth from the triple-logistic model is shown in Figure 8.2 for the participants in the Berkeley Guidance Study; the median value is about $0.15\,cm^2$, corresponding to a standard deviation of the residuals is about 0.4 cm. This value must be very close to the true level of random and short-term variation of such measurements, from such sources as measurement error

Boys	$cm.^2$	Girls
	.5	89
441	.5	
	.4	9
31111	.4	0
9875	.3	59
4321	.3	11122344
96	.2	67799
42210000	.2	00111133
999886666655	.1	66678
4443333211100000	.1	0001112223333444
99887776	.0	666777888889999
	.0	234445

FIGURE 8.2. Back-to-back stem-and-leaf display of the mean square residuals from the triple-logistic model for the data for the Berkeley Guidance Study.

and diurnal and seasonal variation (Goldstein, 1979, p. 38), which are not included in the functional part of this long-term model.

1.2. Some Uses of Fitted Growth Curves

A common use of growth curve fitting has been the comparison of patterns of growth between existing or experimental groups of subjects. There are obvious difficulties in making a succinct comparison of patterns of growth based directly on 10 to 50 observations from birth to maturity. But if the observations for each case are fitted with a function depending on only a few parameters, then the parameters can serve as a multivariate summary of the pattern of growth in each case, and comparisons among the patterns of growth may be made by multivariate analysis of the parameter estimates.

Growth patterns of boys and girls have been compared in this way by Preece and Baines (1978, models 1, 2, and 3) and Largo et al. (1978, smoothing splines) among others. The double-logistic model (Bock et al., 1973) has also been used to compare growth between children of native and European ancestry in Guatemala (Johnston et al., 1976), between normal children and those with Down's syndrome (Rarick et al., 1975), and among the four major growth studies in the United States (Thissen et al., 1976). Summaries of growth and development, such as Malina's (1978) chapter in the *Yearbook of physical anthropology,* use average fitted growth curve parameters to summarize growth data from many different sources.

Fitted growth curve parameters have also been used as data for analysis of hereditary factors in growth and development. Johnston et al. (1976), using the division of growth under the double-logistic model into prepubertal and adolescent components, found that environmental factors seemed to predominate in individual differences in prepubertal growth, whereas hereditary factors appeared more important in adolescent development. Vanderberg and Falkner (1965) and Welch (1970) used the parameters of polynomials in age to evaluate the relative contributions of hereditary and environmental factors to early childhood growth, over a sufficiently short range of age for the polynomials to be a useful representation.

2. Incorporating Information from the Population

In a seminal paper, Lindley and Smith (1972) proposed what they called "Bayes estimation for the linear model." Their ideas were soon applied to problems of fitting growth curves by Rao (1975) and Fearn (1975), as well as in a wide variety of other situations in which parameters are to be estimated for individual units, but those individual units can be assumed to be a sample from some population.

There are two motivations for incorporating information from the population level in the estimation of individual-level parameters. A practical motivation is to solve what Rubin (1980) termed the "bouncing beta" problem. In a situation in which the individual-level parameters were regression weights ("betas"), Rubin (1980) noted that they seemed to bounce excessively between replications, producing unacceptable outlying values. In the "LSAT validation study" Rubin showed that empirical Bayes estimation can suppress such bouncing parameter estimates, because the estimates shrink toward the population mean and are thus less variable across replications. Further, the empirical Bayes estimates can perform better on cross-validation than the least squares estimates, even though the latter minimize squared residuals for each individual set of data.

Reduction of excessive random variation in parameter estimates also motivated Berkey's (1982b) Bayesian approach to individual parameter estimation for the Jenns model and, in part, our work with two-level estimation for the triple-logistic model (Bock and Thissen, 1980). Another motivation for two-level estimation in the growth curve problem is that, as alluded to, it provides efficient estimation of population parameters when the case-level observations are limited, as is often the case in longitudinal studies. In comparative studies of growth in different populations, a multilevel system provides direct estimates of the parameters of the distribution of growth parameters in the population (e.g., their mean and covariance matrix). Such estimates are generally superior to those obtained by averaging individual parameter estimates.

In the two-level approach, in place of averaging, we use Bayes' rule. To do so in this context does not require the specification of a subjective prior distribution and is therefore not controversial on that point. Rather, it merely assumes a joint multivariate normal distribution of the data and the case-level parameters in the population of cases, and employs the mean or mode of the conditional distribution, given the data, as an estimator of the parameters. But by convention, the population distribution of the case-level parameters is called the *prior* distribution, and we will continue that practice here. Similarly, the mean and covariance matrix of the prior distribution are called the *hyperparameters,* a term we will use also.

Empirical Bayes or two-level estimation in linear models has been well developed for some time: Rao (1975) and Fearn (1975) described procedures involving estimators that are linear combinations (or weighted averages) of the case-level least squares estimator and the population mean. The form of this estimator is essentially a multivariate version of Kelley's (1927) regressed estimate used in psychological test theory (see Bock, 1983).

The corresponding problems in multilevel estimation for nonlinear models are still the subject of active research. Laird and Ware (1982) and Racine-Poon (1985) have used the EM algorithm (Dempster, Laird, and Rubin, 1977) to compute multilevel estimates for such models.

In most of the work in growth curve fitting, the residuals are assumed to be independent. If the measurements are sufficiently well separated in time, that assumption might be plausible, but such widely separated time points would not provide sufficient information to fit a complex curve. If the measurements are spaced more closely together in time, some autocorrelation among the residuals must be expected because the error process is not instantaneous. The effect of a random illness, for example, that suppresses growth at time t_i will still have some effect after some small interval, at time t_{i+1}. If this source of autocorrelation is ignored and all observations treated as independent, the fitted curve will be biased toward regions of the data where the observations are more closely spaced. As described here, Bock and Thissen (1980) avoid this difficulty by using spectral analysis to estimate the error autocorrelation function in a large sample of longitudinal data for boys and girls in the Berkeley Guidance Study (Tuddenham and Snyder, 1954). From these functions in real time, the autocorrelation of errors between observations at any arbitrary age intervals can be approximated and employed in the computation of the inverse error matrix that weights the observations properly in a linear or nonlinear estimation procedure.

There has also been some recent work on the simultaneous estimation of residual autocorrelation parameters with growth curve fitting; for examples see Glasby (1979) and Sandland and McGilchrist (1979).

2.1. Multilevel Analysis as Measurement

In this section, we develop the concepts of multilevel analysis as they apply to growth curves. We follow the theoretical development by Bock (1989), who conceives of measurement of individual differences as "inferring the location of the individual in the attribute distribution of the population" (Bock, 1989). We suppose that the probability that the vector observation y on individual i takes on the value y_i (in the case of stature, the longitudinal set of measurements) given the individual's attribute vector θ_i, and the property vector of the measurement operation ζ, is

$$P(Y = y_i \mid \theta_i, \zeta) = f(y_i; \theta_i, \zeta).$$

Since we are discussing growth curves, the individual's attribute vector, θ_i, includes the parameters of the growth curve, and property vector of the measurement operation, ζ, includes the parameters of the distribution of the residuals.

We further suppose that the distribution of the growth curve parameters in the population is known to have density

$$g(\theta; \eta).$$

In our work with the triple-logistic model for human growth in stature, we assume that $g(\theta; \eta)$ is multivariate normal, in which case the hyperparameters, η, include the mean and covariance matrix for the population distribution of the growth curve parameters.

It then follows from Bayes rule that the "posterior density" of θ (the growth curve parameters) given the observed data y_i is

$$p(\theta \mid y_i) = \frac{f(y_i; \theta_i, \zeta)g(\theta; \eta)}{h(y_i)}, \tag{1}$$

in which

$$h(y_i) = \int_\theta f(y_i; \theta_i, \zeta)g(\theta; \eta)\, d\theta$$

is the marginal probability of the observation, the integral being taken over the range of θ. In the terminology of Bayesian statistics, $f(\cdot)$ is called the *likelihood of the parameters given the data*. The product of the *prior* density, $g(\cdot)$, and the *likelihood*, divided by the *marginal* density, is the *posterior* density.

The traditional goal in fitting growth curves is to estimate the parameters, collected here in the vector θ_i, for each case. Redefining this goal as that of inferring the likely values of the growth curve parameters given the observed growth data, we summarize the posterior density in equation (1) by the expected value of the growth curve parameters for individual i:

$$\bar{\theta}_i = \int_\theta \theta p(\theta \mid y_i)\, d\theta,$$

which is called the *Bayes* or (*expected a posteriori*) (EAP) *estimator*, and its posterior covariance matrix

$$\Sigma_{\theta_i|y_i} = \int_\theta (\theta - \bar{\theta}_i)(\theta - \bar{\theta}_i)'p(\theta \mid y_i) \, d\theta.$$

In general, $\bar{\theta}_i$ is a biased estimate of θ_i; but it is nonetheless optimal in terms of the average mean square error for the population. Despite this optimality, the practical usefulness of the posterior mean in the context of growth curve models is limited by the computational barrier of high-dimensional numerical integration, which cannot be easily overcome even by high-speed computers.

In our work with the triple-logistic model, we have therefore made use of the mode, rather than the mean, of the posterior density. The mode is computed by solving the (*maximum a posteriori*) (MAP) equations to locate $\hat{\theta}_i$, the value of θ at which $p(\theta \mid y_i)$ has its maximum value, where

$$\frac{\partial \log p(\theta \mid y_i)}{\partial \theta} = \left[\frac{\partial \log f(y_i; \theta, \zeta)}{\partial \theta}\right]_{\theta = \hat{\theta}_i} + \left[\frac{\partial \log g(\theta; \eta)}{\partial \theta}\right]_{\theta = \hat{\theta}_i} = 0. \quad (2)$$

To solve these nonlinear equations, we use Newton–Raphson iterations, with the matrix of second derivatives of the log posterior density

$$\frac{\partial^2 \log p(\theta \mid y_i)}{\partial \theta \, \partial \theta'} = \left[\frac{\partial^2 \log f(y_i; \theta, \zeta)}{\partial \theta \, \partial \theta'}\right]_{\theta = \hat{\theta}_i} + \left[\frac{\partial^2 \log g(\theta; \eta)}{\partial \theta \, \partial \theta'}\right]_{\theta = \hat{\theta}_i},$$

or Fisher's scoring algorithm, using the expected value of this matrix,

$$I(\theta_i) = -\mathcal{E}\left[\frac{\partial^2 \log p(\theta \mid y_i)}{\partial \theta \, \partial \theta'}\right]_{\theta = \hat{\theta}_i}.$$

In the latter case, the correction at each iteration k is given by

$$\hat{\theta}_{k+1} = \hat{\theta}_k + I^{-1}(\hat{\theta}_k)G(\hat{\theta}_k),$$

where

$$G(\hat{\theta}_k) = \frac{\partial \log p(\hat{\theta}_k)}{\partial \theta}.$$

To the extent that the posterior density is multivariate normal, the MAP and EAP estimates coincide. In the context of MAP estimation, the covariance matrix of the posterior distribution is approximated by

$$\Sigma_{\theta_i|y_i} \cong I^{-1}(\theta_i).$$

For the triple-logistic growth curve model,

$$\theta_i = [a_1, b_1, c_1, a_2, b_2, c_2, a_3, b_3, c_3,]_i$$

with the empirically derived constraint that

$$\frac{a_1}{a_1 + a_2} = 0.851 + 0.041b_1 - 0.018c_1 - 0.517b_2 - 0.042c_2. \qquad (3)$$

The motivation for this constraint is discussed further in Section 2.2. Under the constraint, there are eight free parameters to be estimated for each individual. We assume that the residuals have a multivariate normal distribution with mean zero and covariance matrix Σ, where Σ includes the autoregression of the residuals. We further assume that the distribution of the parameters θ, $g(\theta; \eta)$, is multivariate normal with mean μ_θ and covariance matrix Σ_θ. Then the logarithm of the posterior density [from equation (1)] is proportional to

$$\log p(\theta \mid y_i) \sim -\tfrac{1}{2}[y_i - \hat{y}_i]'\Sigma^{-1}[y_i - \hat{y}_i] - \tfrac{1}{2}[\theta_i - \mu_\theta]'\Sigma_\theta^{-1}[\theta_i - \mu_\theta], \qquad (4)$$

and the MAP equations (2) that must be solved are

$$\frac{\partial \log p(\theta \mid y_i)}{\partial \theta} = [y_i - \hat{y}_i]'\Sigma^{-1}\left[\frac{\partial \hat{y}}{\partial \theta}\right]_{\theta = \hat\theta} - \Sigma_\theta^{-1}[\theta_i - \mu_\theta]_{\theta = \hat\theta} = 0. \qquad (5)$$

Bock and Thissen (1980), Bock (1986), and Thissen and Sykes (1984) have discussed the usefulness of estimates computed as the solutions of equation (5) for the characterization of growth and for the prediction of adult stature from limited childhood data. Berkey (1982b) has used a similar system to estimate the parameters of the Jenns curve for data from early childhood growth (using the simplification that $\Sigma = \sigma^2 I$), and reports improvement in performance over the least squares estimates.

2.2. Group-Level Parameters

The MAP equations assume that the error covariance matrix Σ, the elements of which we have represented by the vector ζ, the population mean μ, and covariance matrix Σ_θ, represented by the vector η, are known, when in fact they are not. If our two-level analysis were subjective Bayes, we would simply specify values for the hyperparameters in η and obtain the Bayes estimates of θ. However, the point of the empirical Bayes procedure is to estimate the values of ζ and η using the observed data y_i from the cases under study.

In principle, the hyperparameters, η, may be estimated by the MML method by integrating over the growth parameter distribution and maximizing the likelihood of the entire sample of N cases:

$$L_M = \prod_{i=1}^{N} h(y_i).$$

In practice, MML estimation of the hyperparameters has not yet been attempted in nonlinear models, where no closed-form solutions are available and the dimensionality is too high for current techniques of numerical integration (except perhaps Monte Carlo integration). An alternative is to approximate the full MML solution by the Dempster, Rubin, and Tsutakawa (1981) EM procedure for linear models, by substituting the posterior mode for the posterior mean, and substituting the inverse posterior information matrix for the posterior covariance matrix. The TRIFIT program (Bock and Thissen, 1983) incorporates such an EM step. We are continuing to study the problems for numerical analysis presented by the nonlinear case.

Careful attention to numerical analysis is also necessary for the full solution of the MAP equations

$$\frac{\partial \log L_M}{\partial \zeta} = \sum_{i}^{N} \frac{1}{h(y_i)} \left[\frac{\partial h(y_i; \zeta, \eta)}{\partial \zeta} \right]_{\substack{\zeta = \hat{\zeta} \\ \eta = \hat{\eta}}} = 0 \qquad (6)$$

and

$$\frac{\partial \log L_M}{\partial \eta} = \sum_{i}^{N} \frac{1}{h(y_i)} \left[\frac{\partial h(y_i; \zeta, \eta)}{\partial \eta} \right]_{\substack{\zeta = \hat{\zeta} \\ \eta = \hat{\eta}}} = 0. \qquad (7)$$

To estimate the autocorrelation structure of the residuals from the fitted growth curves, we assume that the errors of measurement are distributed multinormally with mean zero and covariance matrix $\Sigma = \sigma^2 R$, where R is an autocorrelation matrix in which elements in each diagonal are equal. In Bock and Thissen (1980), we estimated Σ following, in part, a suggestion by Fearn (1975). First, we computed provisional residuals from the triple-logistic model for 67 boys and 62 girls of the Berkeley Guidance Study (Tuddenham and Snyder, 1954) by fitting the model with nonlinear least squares.[2] We then calculated the covariance matrix of the residuals as a provisional estimate of Σ and averaged the diagonal elements to obtain an estimate of σ^2.

---

[2] We attempted to fit 70 boys and 64 girls. But, as is usually the case, a few of the individuals who exhibited "outlying" parameters were eliminated from the residual analysis.

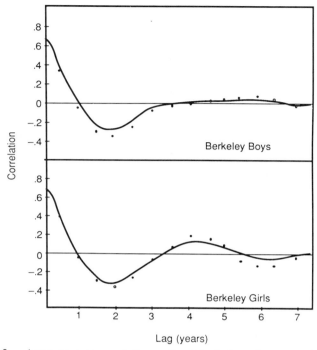

Lag (years)

FIGURE 8.3.   Average autocorrelations of residuals from least squares fits of the data for the Berkeley Guidance participants. The solid line is the autocorrelation function described in the text (from Bock and Thissen, 1980).

Dividing the off-diagonal elements by σ^2 to obtain correlations, we found as expected that the correlations between the residuals were nonzero for relatively close ages and appeared to to depend only on the lag $t_k - t_j$ between measurements k and j. Figure 8.3 shows the average autocorrelation between residuals at various lags, in years, for six-month intervals. They show the typical pattern of an autoregressive process of order greater than 1 (Box and Jenkins, 1976, pp. 53–66).[3]

The solid curves in Figure 8.3 were obtained by fitting a function of the form

$$\rho(k) = 2\left[0.7 + \sum_{i=1}^{n_f} g_i \cos\left(\frac{\pi k_i}{n_f}\right)\right], \qquad (8)$$

in which $\rho(k)$ is the correlation for lag k (in months), n_f is the number of points considered (in this case, 40), and g_i represents the Fourier weights.

[3] See Chapter 9 for an extensive discussion of spectral analysis.

The value 0.7 in equation (8) represents the maximum autocorrelation; 0.3 is the proportion of the residual variance attributed to measurement error.

Using the autoregressive estimates of the covariances of the residuals, we may compute an estimate of Σ for the measurements among arbitrarily time-spaced points. Given this value for Σ, we estimate the θ_i, μ_θ, and Σ_θ. Although it may be optimal to attempt MML estimation of some restricted parametric representation of Σ, that has not been practical up to this time [but see Glasby (1979) and Sandland and McGilchrist (1979) for work on the estimation of the covariance of residuals]. Because we are interested primarily in the estimates of θ_i, μ_θ, and Σ_θ, and because the values in Σ serve only as "weights" in equation (5) for the estimation of the θ_i, it appears that even a rough approximation of Σ serves our purpose quite well.

As an estimate of μ_θ and Σ_θ, Berkey (1982b) used the mean and covariance matrix of the residuals from least squares curve fitting to obtain the "Bayes" estimates for the Jenns model. In our work with the triple-logistic model, we have thus far only attempted the EM solution for linear models mentioned here.

When we applied this method with no constraints on the nine

TABLE 8.2. Estimates of the Population Mean (μ_θ) for the Triple-Logistic Growth Curve for Boys and Girls of the Berkeley Guidance Study

Parameter	Boys ($N = 66$)		Girls ($N = 70$)	
	Mean	S.D.	Mean	S.D.
$a_1 + a_2$	155.32	5.45	147.82	7.53
b_1	0.82	0.52	1.23	0.76
c_1	−0.47	0.46	−0.24	0.44
b_2	0.41	0.06	0.44	0.12
c_2	7.15	1.41	5.69	1.67
a_3	25.28	4.31	19.39	5.17
b_3	1.14	0.22	1.27	0.24
c_3	13.75	1.07	11.64	0.87

The parameter $a_1 + a_2$ is estimated, and resolved into a_1 and a_2 using the linear relation in equation (3).
From Bock and Thissen, 1980.

parameters of the model, Σ_θ became essentially singular after two or three cycles, indicating some degree of linear dependency among the parameters. This occurs because there are more parameters in the model than there are dimensions of individual difference variation. Pivotal reduction of the estimated matrix led to the linear constraint noted in equation (3). Essentially the same constraint was found in both the boys' and girls' data. Even with this constraint, the estimated Σ_θ is very poorly conditioned; we are investigating the possibility that a further constraint, different for the two sexes, could be imposed without an adverse effect on the goodness-of-fit of the growth curves.

The values of μ_θ obtained for 66 boys and 70 girls of the Berkeley Guidance Study are shown in Table 8.2; these are the parameters for the "average" growth curve shown in Figure 8.1. We use these estimates, enhanced by further computations with data from the Fels sample, in our description of individual curves and and for height prediction as described in the following section.

3. Nonlinear Prediction

We find that the triple-logistic model fits individual growth data for stature rather well. Figure 8.4 (from Bock, 1986) illustrates the fit for one of the participants in the Fels longitudinal study. The open circles are the observed data points, and the heavy solid line is the fitted growth curve. We note that, although data are missing for the ages between 10 and 13, the fit is not impaired. The light solid, dashed, and dotted lines are the cross-sectional 5th, 25th, 50th, 75th, and 95th percentiles of growth from the Ross Laboratory charts based on the Fels data. Notice that the growth curve for this child tracks the cross-sectional norms, except at puberty. The cross-sectional norms have the property Merrell (1931) described for such curves: differences in the timing of puberty between individuals cause average growth to appear "flatter" during the growth spurt than does any individual curve. The solid triangles in the lower part of the graph are observed annual growth velocities, and the dashed curve is the derived velocity from the growth curve.

The closed circles in Figure 8.4 plot skeletal maturity against age (using the velocity scale as years instead of centimeters); the horizontal line crosses the skeletal age line at approximately the age of peak height velocity in adolescence for boys. Although the age at peak height velocity during the adolescent growth spurt exhibits substantial individual

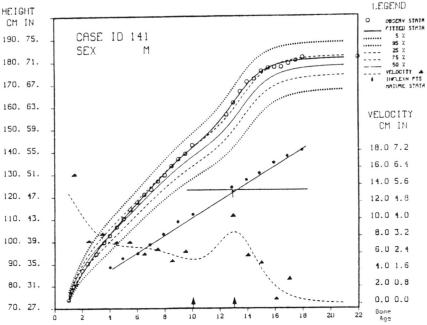

FIGURE 8.4. The fit for one of the participants in the Fels longitudinal study
(#141). The open circles are the observed data points, and the heavy solid line is
the fitted growth curve. The light solid, dashed and dotted lines are the
cross-sectional 5th, 25th, 50th, 75th, and 95th percentiles of growth, showing the
way the individual curve "tracks" the cross-sectional norms, except at puberty.
The solid triangles in the lower part of the graph are observed annual growth
velocities, and the dashed curve is the derived velocity from the growth curve.
The closed circles plot skeletal maturity (using the velocity scale as years instead
of cm) against age; the horizontal line crosses the skeletal age line at
approximately the age of peak height velocity in adolescence for boys. (Reprinted
with permission from Taylor and Francis, Bock, R. D. "Unusual growth patterns
in the Fels data" in A. Demirjian, ed., *Human growth: multidisciplinary review*,
p. 73 (1986).)

differences, the skeletal age (estimated from hand-wrist radiograms) at
peak height velocity shows much less variation. We can use this fact to
improve prediction of mature stature as discussed later.

The prediction of mature stature from childhood growth data is a
subject of intense interest for some parents, and for pediatricians
involved in the diagnosis and treatment of possible disorders of growth.
Data from the major longitudinal studies of growth have played an
important part in the development of systems for such prediction.

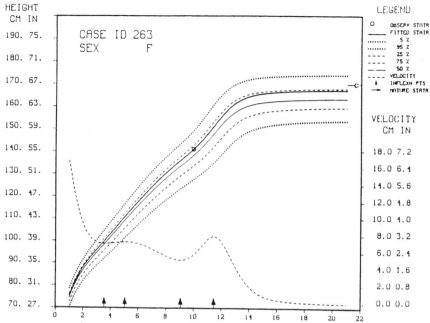

FIGURE 8.5. Shows the triple-logistic fitted to a single measurement (at age 10) for another of the participants (#263) in the Fels longitudinal study. The open circle is the observed data point, and the heavy solid line is the fitted growth curve. The light solid, dashed and dotted lines are the cross-sectional 5th, 25th, 50th, 75th, and 95th percentiles of growth. The fitted curve reaches an asymptote just below her actual mature stature, shown by the horizontal arrow inside the plot frame at the upper right. (Reprinted with permission from Taylor and Francis, Bock, R. D. "Unusual growth patterns in the Fels data" in A. Demirjian, ed., *Human growth: multidisciplinary review*, p. 82 (1986).)

Analysts who have worked with these data have pointed out that empirical Bayes methods of curve fitting provide a flexible system of prediction for this purpose (Fearn, 1975; Bock and Thissen, 1980; Racine-Poon, 1985; Bock, 1986). When the parameters of the individual growth curve are computed as the solution to the MAP equations (5), the growth curve can be estimated from any number of data points, including a single data point.[4] Figure 8.5 shows the triple-logistic curve fitted to a single measurement at age 10 for another of the participants in the Fels longitudinal study. The fitted curve reaches as asymptote just below her

---

[4] There is even a solution to equation (5) for no data: it is the population mean curve. As the number of observations for the case increases, the fitted curve is pulled away from the population mean curve to the true curve for that person.

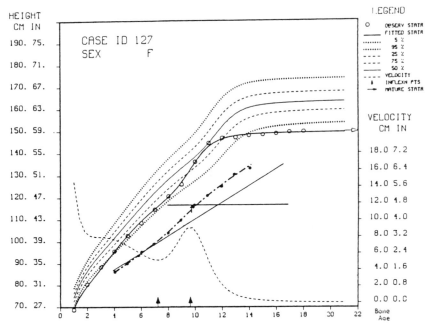

FIGURE 8.6. The triple-logistic for another of the participants (#127) in the Fels longitudinal study. The open circles are the observed data points, and the heavy solid line is the fitted growth curve. The light solid, dashed and dotted lines are the cross-sectional 5th, 25th, 50th, 75th, and 95th percentiles of growth. The closed circles plot skeletal maturity (using the velocity scale as years instead of cm) against age; the horizontal line crosses the skeletal age line at approximately the age of peak height velocity in adolescence for girls. (Reprinted with permission from Taylor and Francis, Bock, R. D. "Unusual growth patterns in the Fels data" in A. Demirjian, ed., *Human growth: multidisciplinary review*, p. 80 (1986).)

actual mature stature, shown by the horizontal arrow inside the plot frame at the upper right; the prediction is within 2 cm of the correct value. The mature height of about 90% of the Fels cases can be predicted from age 10 in girls and age 12 in boys with this accuracy (Bock, 1986).

Among the remaining 10%, however, are unusual cases like Fels case #127, the shortest girl in the Fels sample. Her data and fitted growth curve are shown in Figure 8.6. She is short throughout childhood, has a relatively early adolescent growth spurt, and grows very little after about age 12. Her skeletal age is increasingly more mature than her chronologi-cal age after about age 8, indicating very rapid maturation. Figure 8.7 illustrates the attempt to predict adult stature for this case from a

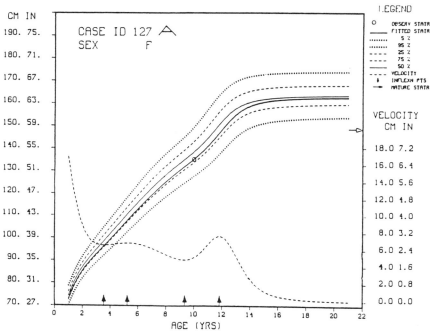

FIGURE 8.7. Shows the triple-logistic fitted to a single measurement (at age 10) for participant #127 in the Fels longitudinal study. The open circle is the observed data point, and the heavy solid line is the fitted growth curve. The light solid, dashed and dotted lines are the cross-sectional 5th, 25th, 50th, 75th, and 95th percentiles of growth. The fitted curve reaches an asymptote well above her actual mature stature, shown by the horizontal arrow inside the plot frame at the upper right. (Reprinted with permission from Taylor and Francis, Bock, R. D. "Unusual growth patterns in the Fels data" in A. Demirjian, ed., *Human growth: multidisciplinary review*, p. 82 (1986).)

single observation at age 10. It happens that, because near age 10 she is of nearly average stature, a single observation does not pull the estimated growth curve parameters far from the population mean, and a near average adult stature is predicted with an error of about 11 cm.

There are two ways to improve upon the situation. One is to use skeletal maturity, in addition to stature, as a predictor at age 10. Skeletal age is commonly used in linear prediction systems to distinguish between those cases who are maturing early and late (Roche, Wainer, and Thissen, 1975; Tanner et al., 1975). To improve prediction with the triple-logistic model, we note that there is very little variance in the

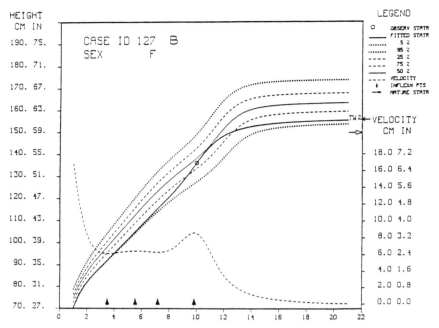

FIGURE 8.8.   Shows the triple-logistic fitted to a single measurement (at age 10) for participant #127 in the Fels longitudinal study, with the location of the adolescent component constrained to fall before age 10. The open circle is the observed data point, and the heavy solid line is the fitted growth curve. The light solid, dashed and dotted lines are the cross-sectional 5th, 25th, 50th, 75th, and 95th percentiles of growth. The fitted curve reaches an asymptote about 6 cm above her actual mature stature, shown by the horizontal arrow inside the plot frame at the upper right. (Reprinted with permission from Taylor and Francis, Bock, R. D. "Unusual growth patterns in the Fels data" in A. Demirjian, ed., *Human growth: multidisciplinary review*, p. 83 (1986).)

skeletal age at which pubertal peak height velocity occurs. As is clear in the skeletal ages plotted in Figure 8.6, #127 had a skeletal age well beyond the average skeletal age at peak height velocity when she was 10. Therefore, we convey this information to the height prediction system by changing μ_θ and Σ_θ so that the parameter c_3, which controls the age of adolescent peak height velocity, has a high prior density before age 10. With this added information, the fitted growth curve appears as in Figure 8.8. The predicted curve based on a single observation, and the information that her skeletal age at 10 was about 12, is now much closer to the true curve.

Even greater improvements may be made over the originally poor prediction for #127 if more data are used in order to provide information

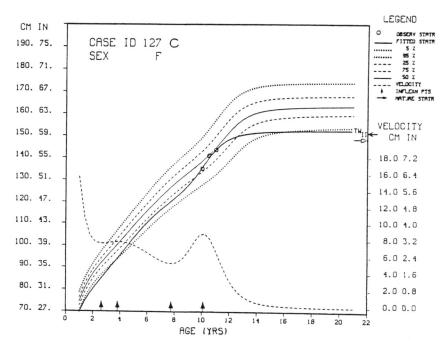

FIGURE 8.9. Shows the triple-logistic fitted to three measurements (at age 10–11) for a participant (#127) in the Fels longitudinal study. The open circles are the observed data points, and the heavy solid line is the fitted growth curve. The light solid, dashed and dotted lines are the cross-sectional 5th, 25th, 50th, 75th, and 95th percentiles of growth. The fitted curve reaches as asymptote within about 2 cm of her actual mature stature, shown by the horizontal arrow inside the plot frame at the upper right. (Reprinted with permission from Taylor and Francis, Bock, R. D. "Unusual growth patterns in the Fels data" in A. Demirjian, ed., *Human growth: multidisciplinary review,* p. 83 (1986).)

about growth velocity. Figure 8.9 shows the fitted curve using three observations between ages 10 and 11 in place of the single observation at 10. The data are now sufficient to convey the fact that growth is decelerating and to determine a Bayes estimated curve that is within 2 cm of the correct adult stature.

The virtue of this Bayesian method of growth-curve prediction is that it can use almost any kind or amount of collateral information to improve its performance by the same mechanism used for skeletal age above—by "conditioning" the values of μ_θ and Σ_θ on other variables such as weight and parental stature. In addition, any number of observations of height may be included. In contrast, linear height prediction systems, which must be "calibrated" for each variable to be included, are very difficult to change in order to add some other variable if that seems, after the fact, to be useful. The RWT height prediction system (Roche, Wainer, and Thissen, 1975), for example, includes a single measurement of current stature, parental stature, weight, and skeletal age. If another observation of childhood stature is available, only one can be used because the linear equation has not been calibrated for two. Similarly, the Tanner–Whitehouse, Mark 3, linear prediction system can accept two observatons, but not three.

4. Conclusions

In the more than 50 years since Merrell (1931) made it clear that individually fitted growth curves were necessary in the analysis of growth data, improved technology has put advanced statistical procedures for this purpose at the disposal of anyone who has access to a computer. The recent introduction of multilevel statistical methods has opened the way to applications of this technology in the dependable fitting of both linear and nonlinear structural models to growth data. Although we have discussed only applications to growth in height, nothing prevents similar models and methods from being applied to measures of cognitive and affective development, provided modern psychometric methods are used to place the data on a scale with well-defined units.[5]

Although the extension of growth-curve fitting to other variables will presumably not require another half century, we can expect that it will follow roughly the same steps as the work on human stature. First, on some theoretical and empirical basis, an appropriate family of curves must be specified. In the beginning, extensive and high-quality data will be needed for at least a small sample of cases in order to reveal something of the range of individual differences that the model must accommodate and to estimate the common error structures for residuals.

After some plausible models have been specified and found by ordinary least squares methods to fit acceptably in the highly detailed

---

[5] The traditional summed scores (of the number of positive responses) on psychological tests are clearly not good candidates for growth-curve fitting, because there is no reason to believe that such scores represent equal-interval measurement or that they should follow any particular mathematical function of age. However, the scaled scores produced by applications of modern psychometric theory (item response theory) have well-defined units and may be used; Bock (1975, pp. 453ff) illustrates the use of such scores.

data, large samples of cases can be collected, possibly with more sparse measurement, but concentrating on those intervals of the growth curve that are most informative for the models in question. From these large samples, the population distributions of the parameters for individual growth can be estimated by marginal maximum likelihood and incorporated in the empirical Bayes procedures for fitting at the case level. Finally, if prediction of growth is required, the relationship of other variables to the parameters of the growth model can be investigated by multivariate statistical methods in order to find collateral information that improves prediction. If sufficient longitudinal data are available to evaluate the results, a formal system of prediction with specified accuracy can then be incorporated in the Bayes fitting program for general use.

Advances in fitting growth curves are not complete, by any means. As we have seen, approximations are still required at many stages, especially in estimating the population distributions efficiently. We expect, however, that the next generation of parallel-processing computers will be much better suited to numerical integration than are current machines and will allow the subject matter, rather than the limitations of our computational capabilities, to determine our procedures for the quantitative analysis of growth and development.

References

Berkey, C. S. (1982a). "Comparison of two longitudinal growth models for pre-school children," *Biometrics.* **38,** 221–234.

Berkey, C. S. (1982b). "Bayesian approach for a nonlinear growth model," *Biometrics.* **38,** 953–961.

Bock, R. D. (1975). *Multivariate statistical methods in behavioral research.* New York: McGraw-Hill.

Bock, R. D. (1983). "The discrete Bayesian." In H. Wainer and S. Messick (Eds.), *Principals of modern psychological measurement* (pp. 103–115). Hillsdale, N.J.: Lawrence Erlbaum.

Bock, R. D. (1986). "Unusual growth patterns in the Fels data". In A. Demirjian (Ed.), *Human growth: multidisciplinary review.* London and Philadelphia: Taylor and Francis.

Bock, R. D. (1989). Measurement of human variation: A two-stage model. In R. D. Bock (Ed.), *Multilevel analysis of educational data* (pp. 319–342). San Diego, Academic Press.

Bock, R. D., and Thissen, D. (1976). "Fitting multi-component models for growth in stature," *Proceedings of the 9th International Biometric Conference,* **1,** 431–442.

Bock, R. D., and Thissen, D. (1980). "Statistical problems of fitting individual growth curves." In F. E. Johnston, A. F. Roche, and C. Susanne (Eds.), *Human physical growth and maturation: methodologies and factors* (pp. 265–290). New York: Plenum.

Bock, R. D., and Thissen, D. (1983). *TRIFIT: A program for fitting and displaying triple logistic growth curves.* Mooresville, Ind.: Scientific Software.

Bock, R. D., Wainer, H., Petersen, A., Thissen, D., Murray, J., and Roche, A. (1973). "A parameterization for individual human growth curves," *Human Biology.* **45,** 63–80.

Box, G. E. P., and Jenkins, G. M. (1976). *Time series analysis: forecasting and control.* San Francisco: Holden-Day.

Burt, C. (1937). *The backward child.* New York: Appleton-Century.

Chandler, P., and Bock, R. D. (1988). Analysis of age changes in adult stature from longitudinal data. Unpublished manuscript.

Count, E. W. (1943). "Growth patterns of the human physique: an approach to kinetic anthropometry," *Human Biology.* **15,** 1–32.

Dempster, A. P., Laird, N. M., and Rubin, D. B. (1977). "Maximum likelihood with incomplete data via the E-M algorithm," *Journal of the Royal Statistical Society, Series B.* **39,** 1–38.

Dempster, A. P., Rubin, D. B., and Tsutakawa, R. K. (1981). "Estimation in covariance components models," *Journal of the American Statistical Association.* **76,** 341–353.

Dunn, H. G., Hughes, G.-J., and Schulzer, M. (1986). "Physical growth." In H. G. Dunn (Ed.), *Sequelae of low birthweight: the Vancouver study* (pp. 35–53). Clinics in Developmental Medicine, No. 95/96. Mac Keith Press, Oxford: Blackwell.

el Lozy, M. (1978). "A critical analysis of the double and triple logistic growth curves," *Annals of Human Biology.* **5,** 389–394.

Eveleth, P. B., and Tanner, J. M. (1976). *Worldwide variation in human growth.* IBP Synthesis Series 8. Cambridge, England: Cambridge University Press.

Fearn, T. (1975). "A Bayesian approach to growth curves," *Biometrika.* **62,** 89–100.

Frisch, R. E. (1988). "Fatness and fertility," *Scientific American.* **258,** 88–95.

Gasser, T., Müller, H., Köhler, W., Prader, A., Largo, R., and Molinari, L. (1985). "An analysis of the mid-growth and adolescent spurts of height based on acceleration," *Annals of Human Biology.* **12,** 129–148.

Glasby, C. A. (1979). "Correlated residuals in non-linear regression applied to growth data," *Applied Statistics.* **28,** 251–259.

Goldstein, H. (1979). *The design and analysis of longitudinal studies.* New York: Academic Press.

James, W., and Stein, C. (1961). "Estimation with quadratic loss." In *Proceedings of the 4th Berkeley Synposium on Mathematical Statistics and Probability,* Vol. 1, pp. 361–379.

Jenns, R. M., and Bayley, N. (1937). "A mathematical model for studying the growth of a child," *Human Biology.* **9,** 556–563.

Johnston, F. E., Wainer, H., Thissen, D., and MacVean, R. (1976). "Hereditary and environmental determinants of growth in height in a longitudinal sample of children and youth of Guatemalan and European ancestry," *American Journal of Physical Anthropology.* **44,** 469–476.

Kelley, T. L. (1927). *The interpretation of educational measurements.* New York: World Books.

Laird, N. M., and Ware, J. H. (1982). "Random-effects models for longitudinal data," *Biometrics.* **38,** 963–974.

Largo, R. H., Gasser, Th., Prader, A., Stuetzle, W., and Huber, P. J. (1978). "Analysis of the adolescent growth spurt using smoothing spline functions," *Annals of Human Biology.* **5,** 421–434.

Lindley, D. V., and Smith, A. F. M. (1972). "Bayes estimates for the linear model," *Journal of the Royal Statistical Society, Series B.* **34,** 1–42.

Malina, R. M. (1978). "Adolescent growth and maturation: selected aspects of current research," *Yearbook of Physical Anthropology.* **21,** 63–94.

Manwani, A. H., and Agarwal, K. N. (1973). "The growth pattern of Indian infants during the first year of life," *Human Biology.* **45,** 341–349.

Maritz, J. S. (1970). *Empirical Bayes methods.* London: Methuen.

Marubini, E., Resele, L. F., and Barghini, G. (1971). "A comparative fitting of the Gompertz and logistic curves to longitudinal height data during adolescence in girls," *Human Biology.* **43,** 237–252.

Marubini, E., Resele, L. F., Tanner, J. M., and Whitehouse, R. H. (1972). "The fit of Gompertz and logistic curves to longitudinal data during adolescence on height, sitting height and biacromial diameter in boys and girls of the Harpenden Growth Study," *Human Biology.* **44,** 511–524.

Merrell, M. (1931). "The relationship of individual growth to average growth," *Human Biology.* **3,** 37–70.

Mosteller, F., and Tukey, J. W. (1977). *Data analysis and regression.* Reading, Mass.: Addison-Wesley.

Peschel, E. R., and Peschel, R. E. (1987). "Medical insights into the castrati in opera," *American Scientist.* **75,** 578–583.

Preece, M. A. (1986). "Prepubertal and prepubertal endocrinology." In F. Falkner and J. M. Tanner (Eds.), *Human growth,* Vol. 2 (pp. 211–241). New York: Plenum.

Preece, M. A., and Baines, M. J. (1978). "A new family of mathematical models describing the human growth curve," *Annals of Human Biology.* **5,** 1–24.

Racine-Poon, A. (1985). "A Bayesian approach to nonlinear random effects models," *Biometrics.* **41,** 1015–1023.

Rao, C. R. (1975). "Simultaneous estimation of parameters in different linear models and applications to biometric problems," *Biometrics.* **31,** 545–554.

Rarick, G. L., Wainer, H., Thissen, D., and Seefeldt, V. (1975), "A double logistic comparison of growth patterns of normal children and children with Down's syndrome," *Annals of Human Biology.* **2,** 339–346.

Reed, R. B., and Stuart, H. C. (1959). "Patterns of growth in height and weight from birth to 18 years of age," *Pediatrics.* **24,** 904–921.

Robertson, T. B. (1908). "On the normal rate of growth of an individual, and its biochemical significance," *Archiv fur Entwicklungsmechanik der Organismen.* **25,** 581–614.

Roche, A. F., Wainer, H., and Thissen, D. (1975). "Predicting adult stature for individuals," *Monographs in Pediatrics,* Whole No. 3. Basel, Switzerland: Karger.

Rubin, D. B. (1980). "Using empirical Bayes techniques in the Law School validity studies," *Journal of the American Statistical Association.* **75,** 801–816.

Sandland, R. L., and McGilchrist, C. A. (1979). "Stochastic growth curve analysis," *Biometrics.* **35,** 255–271.

Tanner, J. M. (1978). *Foetus into man: physical growth from conception to maturity.* Cambridge, Mass.: Harvard University Press.

Tanner, J. M., Whitehouse, R. H., Marshall, W. A., Healy, M. J. R., and Goldstein, H. (1975). *Assessment of skeletal maturity and prediction of adult height.* London: Academic Press.

Thissen, D., Bock, R. D., Wainer, H., and Roche, A. F. (1976). "Individual growth in stature: a comparison of four growth studies in the U.S.A.," *Annals of Human Biology.* **3,** 529–542.

Thissen, D., and Sykes, R. C. (1984). "Prediction of growth using the triple logistic model." Paper presented at the annual meeting of the Psychometric Society, Santa Barbara, Calif.

Tuddenham, R. D., and Snyder, M. M. (1954). "Physical growth of California boys and girls from birth to 18 years," *University of California Publications in Child Development.* **1,** 183–364.

Vandenberg, S. G., and Falkner, F. (1965). "Hereditary factors in human growth," *Human Biology.* **37,** 357–365.

Welch, Q. B. (1970). "A genetic interpretation of variation in human growth patterns," *Behaviour Genetics.* **1,** 157–167.

Winsor, C. P. (1932). "The Gompertz curve as a growth curve," *Proceedings of the National Academy of Sciences.* **18,** 1–8.

Spectral Analysis of Psychological Data*

Chapter 9

RANDY J. LARSEN

Department of Psychology
The University of Michigan
Ann Arbor, Michigan

Abstract

Spectral analysis is described as a technique that examines a sequence of observations for underlying rhythms or cycles. This technique "discovers" rhythms in data by applying a series of trigonometric functions—sine and cosine waves—to the data and assessing the degree of fit between the functions and the data. Spectral analysis has the ability to detect several, superimposed rhythms in a given data set, even if such rhythms are hidden by random fluctuations or measurement error. Results of spectral analysis are expressed in terms of variance accounted for by rhythms of different periodicities or frequency. This technique is especially suited for the study of phenomena that are suspected of being rhythmic, where the researcher is interested in detecting rhythmicity or in assessing the period length of one or more rhythms. A worked example applies spectral analysis to the study of mood variability. Computer packages for spectral analysis are discussed.

Many researchers, especially in the area of developmental psychology, observe their subjects over time. If the observations are made inten-

* Preparation of this chapter was supported by Research Scientist Development Award #MHOO704 and grant #MH42057 from the National Institute of Mental Health.

sively, then the measures taken at adjacent time points are unlikely to be independent of each other. For example, suppose the behavior of an infant was observed every 5 minutes for an entire day and rated for the degree of distress exhibited (e.g., crying). We might assume that the infant would exhibit some fluctuation between phases of distress and phases of rest, and that these phases alternate fairly regularly. We might also assume that phases of distress last for several observational occasions and phases of rest last for several observational occasions. In this situation, the current behavior of the infant is dependent, to some degree, on past behaviors. Consequently, prediction of the infant's current state could be based on the recent past states of that infant. In this case the probability that the infant will be crying on a given observational occasion is not independent of time. This example illustrates a case of sequential dependency in a data set, where successive observations are interdependent on each other.

We can contrast this example with the classical probability example of predicting whether a single role of dice will result in two sixes. Here the probability of making this observation is independent of time. The probability of observing two sixes on any given roll of the dice is always the same over time, regardless of how often the dice are rolled and regardless of when in time the dice are rolled. In contrast, the probability of the infant being in a crying or not crying state is dependent on when in time the observation is made and on what state the infant was in on prior occasions of observation. Assuming that the crying phases continue over several occasions of observation before stopping, then there would be a higher probability of finding the infant in a crying state on one occasion if he or she was crying on the prior occasion.

Both the crying behavior of the infant and the roll of the dice are random processes. In both case there are specific probabilities for the various outcomes of observation. These probabilities, however, are of a quite different nature between the two examples. In the dice example, the probabilities on each observation are related to identical, independent distributions. Most statistics used in psychological research consider only such identical, independently distributed probabilities, plus the special-case violations of identical distributions found in linear models and the special-case violations of independence found in repeated-measures analysis of variance.

In the infant distress example, however, both independence and identical distribution assumptions are violated because observations have been made on the same process over sequential time periods. In such a data set, if we correlate pairs of observations taken at different time units apart (i.e., at different time lags), then it is unlikely that these correlations (called the *autocorrelation function*, ACF) will be zero. In fact, the ACF over different time lags will deviate from zero precisely to the extent that the observations *are* sequentially dependent (Brillinger, 1975). This fact, while it highlights the weakness of traditional statistics for describing sequential data, provides the springboard for that unique branch of statistics especially suited for such sequential or time-based data.

1. Time-Series Statistics

Time-series analysis is a category of inferential statistics that contains two general subclasses of models. One class of models bases inferences on autocorrelation information in the sequence of observations. These models are used to characterize a given observation in time as a weighted function of past observations of the same process. A representative of this class of models would be the autoregressive integrated moving-average (ARIMA) models. Models such as these are referred to as estimation in the *time domain* and are often used for forecasting purposes (Box and Jenkins, 1976). The other class of time-series models is referred to as estimation in the *frequency domain*, with spectral analysis representing the major model in this domain. In spectral analysis inferences are made as to the presence or absence of periodic components within the observed time series. The effort in spectral analysis is to estimate the degree to which variation in the observed time series may be accounted for by periodic functions of different frequencies.

Time-domain and frequency-domain models are complementary in the sense that both are based on the same information contained in the observed time series. The two models, however, differ in how they represent this information. Time-domain models express the time series in terms of autoregressive or other time-based parameters. Frequency-domain models express the data in terms of trigonometric functions, and the degree to which such functions account for variation in the observed data. In fact, the transformation from frequency- to time-domain representations is straightforwardly symmetric (Gottman, 1981). Thus the choice of which models to use must be made on the basis of the type of inference the researcher wishes to make (e.g., forecasting future *time* points, or assessing the *frequency* components within the data) and how he or she wishes to represent the information in the observed time series. This chapter will be concerned only with the frequency-domain representation of time-series data through spectral analysis.

2. Spectral Analysis: An Intuitive Description

In general terms, spectral analysis involves accounting for variance in data by fitting a model. A similar goal occurs in traditional bivariate regression, where the attempt is to account for variance in the data by fitting a linear model (a regression line). In simple regression the model is a straight line, where one variable is expressed as a linear function of another variable, as follows:

$$Y = a + b_1 X + e.$$

In more complex forms of regression, the model can be made nonlinear. For example, in polynomial regression one variable is still expressed as a function of another variable, but the function is now nonlinear, as in the following quadratic and cubic polynomial equations.

$$Y = a + b_1 X + b_2 X^2 + e \qquad \text{quadratic equation,}$$
$$y = a + b_1 X + b_2 X^2 + b_3 X^3 + e \qquad \text{cubic equation.}$$

In the quadratic equation, as one variable increases, the other variable increases at a slower rate. In the cubic equation, as one variable increases, the other variable increases at a more complicated rate resulting in an elongated S-shaped relationship. If this pattern were to repeat itself in real data, we would need an even higher-order polynomial to fit the data. The point is that we can build complex polynomial models to account for relationships of very unusual shape. Spectral analysis is similar to regression in the sense that both fit mathematical models to account for variance in the observed data. And in both cases the researcher is interested in how well the model fits the data.

Here the similarity of spectral analysis to bivariate regression must end because there are three major differences between these two approaches to fitting models to data. First, spectral analysis is (in the simple case) univariate, where the same observations of some process are made repeatedly over time. That is, the type of data to which spectral analysis is applied is different than typical regression data; spectral analysis is applied to sequential observations of the same process that result in a time-series data set. Second, the form of the models fit in spectral analysis, though nonlinear, are not simply a collection of polynomial terms from nonlinear regression. And third, in spectral analysis many models are applied to the data and the fit is assessed for each rather than simply fitting one model as in typical regression. Each of these important differences will be discussed in detail in order to describe spectral analysis more completely.

2.1. Time-Series Data

Spectral analysis is applied to a unique type of data. Such data are called *time-series data* and are the result of making successive and sequential observations of the same process over time. Usually the time interval between successive observations is constant, so we measure X at time (t), and then measure X again at time $t + 1$, and again at time $t + 2$, and so on, up to time $t + n$. This observational strategy generates the observed time series $X_t, X_{t+1}, X_{t+2}, \ldots, X_{t+n}$. An example of such data might be brain waves measured at each millisecond, traffic fatalities assessed each month in the United States, infant distress level for every 5-minute period over an 8-hour day, etc.

The point is that time-series data are represented as a string of points or as a series of X-values plotted along a time (t) axis. This is different from the bivariate regression case, where we usually think of an X-plotted-by-Y representation. In time-series analysis, we think of the data using an X-plotted-by-time (t) representation. The following example represents time-series data gathered in a study of daily moods. Here we have one subject who kept a record of his moods on a daily basis for 84 consecutive days. Figure 9.1 shows a plot of this subject's data, where the daily mood score is plotted against a time axis that represents the number of days in the observational period (i.e., three months).

2.2. The Form of Spectral Analysis Models

The second difference between spectral analysis and simple regression is that the models fit in spectral analysis, through nonlinear, are *not* from the class of polynomial models used in regression. Rather, the spectral analysis models are defined by trigonometric functions (Chatfield, 1980). Since the objective of spectral analysis is to see if there are periodic functions in the data, the models fit need to be *truly* periodic. No set of polynomial expressions can be truly periodic without also being infinite.

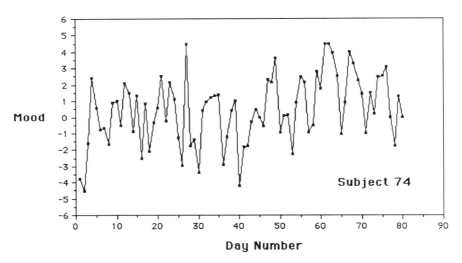

FIGURE 9.1.   Time-series plot of daily mood data from subject 74.

This is so because the collection of polynomial terms needed to produce a truly periodic function would have to contain elements of an infinite order. Thus spectral analysis uses the collection of periodic functions known as sine and cosine functions. An example of a periodic function is the sine wave

$$X_t = A \sin(\omega t + \phi)$$

In this formula, A is the amplitude of the sine wave, ω is its angular frequency (the number of complete cycles made in 2π units of time), and ϕ is the initial phase of the wave. The frequency, represented as the number of cycles per unit time, is expressed as $f = \omega/2\pi$. The *period* of the function (T) is the amount of time from a peak of the wave to the next peak. We can express the period as $T = 2\pi/\omega$. For example, if time is measured in days (or the time interval between successive observations is one day), then the period represents cycles that recur every $2\pi/\omega$ days. The period of a sinusoidal wave is sometimes called its *wavelength*, particularly in engineering applications. Figure 9.2 is an example of a sine wave illustrating the different components of the formula.

    An imaginary example of a sine-wave generator is a pendulum with a pen attached to it swinging over a sheet of paper moving at a constant speed below the pen (Gottman, 1981). The line drawn by the swinging

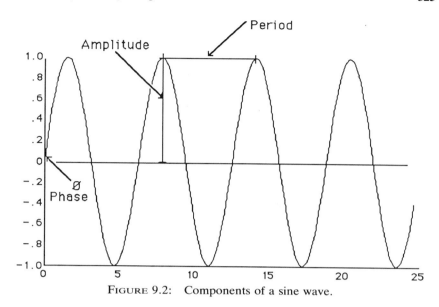

FIGURE 9.2:   Components of a sine wave.

pen would describe a sine wave. The amplitude of that wave is determined by how far the pendulum travels on either side of where it would be if hanging still. If the pendulum were made to swing faster, then it would describe a sine wave of faster frequency $\omega/2\pi$ (and hence a shorter period $2\pi/\omega$). The *phase* of the sine wave refers to when in the arc of its swing we started to make observations or, in general, the direction the periodic process is moving in when we begin observation (i.e., when the pen started writing on the paper).

Thus the second difference between spectral analysis and regression is that spectral analysis fits a *perfectly* repeating, periodic model (in the form of sinusoidal functions) to the data. These waveform functions can be described mathematically with sine-cosine functions. These functions contain parameters referring to the frequency, amplitude, and phase of the periodic cycle.

The given sine-wave equation is a fairly simple model describing a perfect waveform of a single frequency. Real data from the behavioral sciences, as in the example of Figure 9.1, do not look like perfect sine waves. We are hard-pressed to "see" any sine-wave type of variation in the Figure 9.1 time series. However, it is quite possible that this time series may actually contain periodic variation at one or more different frequencies, but this periodic variation may be masked in some way. We will return to this example later. But for now let us think of an example were there might be *two* periodic components in the observed data. For example, traffic fatalities in the Unites States if plotted monthly for several years, might show some interesting periodic variation. We might expect a six-month cycle, with a peak in the winter and a peak in the summer during vacation season. Superimposed on this six-month cycle might be a four-month cycle that is defined by peaks around the holidays that have notoriously high traffic fatalities: New Year's Eve, Memorial Day, and Labor Day. Since these holidays are approximately four months apart, they might generate a sine wave in the traffic fatality data that peaks three times a year (i.e., has a 4-month period, $12/4 = 3$). These two periodic functions—six months and four months—can combine together and, with a little randomness and measurement error thrown in, make it very difficult to "see" either of these periodic components in the data.

The point of the above example is that a single time series may actually contain *several* periodic components. Pure sine waves can be added together, with the result being a complex or irregularly shaped waveform. To illustrate this, imagine the pendulum example again. This time, however, hang two more pendulums from the first so that there are now three weights connected by strings hanging in the form of a compound pendulum. As this compound pendulum swings, it will draw a more complex waveform, one that is the additive function of the three weights and the length of the strings between them. To illustrate this example, imagine three sine waves as in Figure 9.3. Now, if we add these together, we get the more complicated waveform in Figure 9.4. And if we add some random noise into this pattern to simulate real life, we get the pattern of Figure 9.5. The pattern in Figure 9.5 does not look too different from the real daily mood data presented in Figure 9.1. However, we *know* that the data in Figure 9.5 contain three pure sine waves, even though they are difficult to see due to the addition of some random noise. How can be find out if the real daily mood data from Figure 9.1 contain any rhythmic patterns? This is exactly the kind of question that spectral analysis can answer for us. Later we will spectral analyze the data from Figure 9.1 to see if there *are* any periodicities "hidden" in these observations of one subject's daily moods.

The point of this example is that any number of sine or cosine waves may be added together. In addition, random noise might be superimposed on the waveform, somewhat masking the sine waves. Think of the compound pendulum example again, only this time imagine that a mischievous boy with a peashooter has entered the room with the

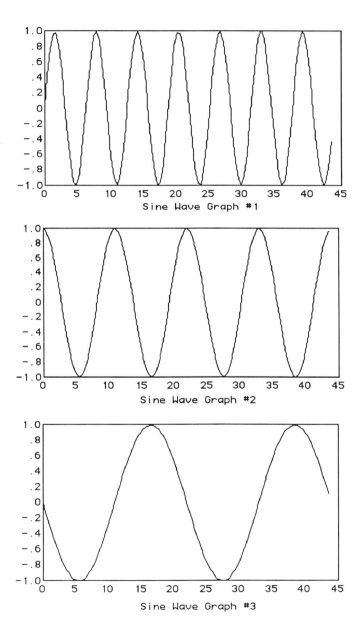

FIGURE 9.3.   Three sine waves that differ in their frequency and phase.

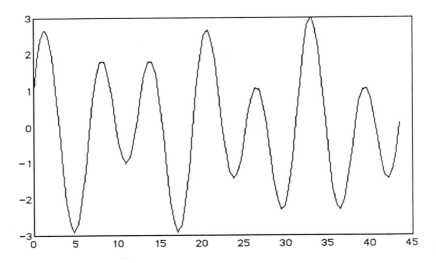

FIGURE 9.4.   Sum of the three sine waves from Figure 9.3.

pendulum (Gottman, 1981). Assume that he is randomly motivated and shoots peas at the swinging pendulum weights and randomly disturbs their motion. These random shocks represent noise that will hide, to some degree, the periodic waveforms generated by the pendulums. From this intuitive example, you see that periodicities may be hidden in an

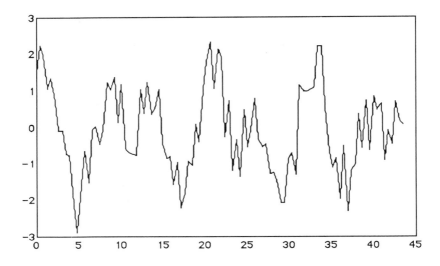

FIGURE 9.5.   Sum of the three sine waves from Figure 9.3 plus random noise.

observed time series due to both noise (random shocks or measurement error) and the presence of two or more periodic components that may differ in their frequency, amplitude, or phase and that combine to generate an irregular waveform.

2.3. Fitting Multiple Functions to the Observed Time-Series Data

The fact that an observed time series may contain several periodic functions leads to the third major difference between regression and spectral analysis. Whereas regression fits a single model to the data (or, more precisely, takes a single estimate of variance accounted for by fitting a model to the data), spectral analysis fits a series or whole spectrum of sinusoidal models to the data and computes a separate estimate of variance accounted for by each of the sinusoidal models. That is, instead of fitting a single sine wave, spectral analysis fits a *collection* of sinusoidal waveform functions

$$X_t = \sum_{j=1}^{n} A_j \sin(\omega_j t + \phi_j)$$

But since $A \sin(\omega t + \phi) = a \cos(\omega t) + b \sin(\omega t)$, where $a = A \sin \phi$, and $b = A \cos \phi$, the equation can be rewritten as

$$X_t = \sum_{j=1}^{n} [a_j \cos(\omega_j t) + b_j \sin(\omega_j t)]$$

which illustrates that the amplitude A and the phase ϕ are functions of the parameters a_j and b_j. So the problem reduces to one of estimating a_j and b_j. It can be shown (Gottman, 1981, p. 204) that the least squares estimates for a_j and b_j are the sample autocovariances of the time series. In fact, estimating a_j and b_j from autocovariances is the objective of Fourier approximation (Bloomfield, 1976).

In Fourier analysis a series of selected periodic components are fit to the time series. These components are selected because they define an overtone series, which are harmonics of one fundamental frequency. The frequency for any time series is $(T - 1)/2$, where T equals the total number of observations in the time series, with overtone series periods of $T, T/2, T/3, \ldots$ to a period of only two time units in length. In other words, the angular frequencies (ω_j) for use in the foregoing equations are determined by $2\pi j/T$, where $j = 1, 2, 3, \ldots, (T - 1)/2$. Once T is given, the overtone series is determined and the amount of periodic variance at the frequencies defined by the overtone series is estimated using the

Fourier transform of the autocovariances of the series. This is a computation-intensive process. It turns out, however, that it is possible to reduce the computations dramatically if T is not a prime number. In fact, the closer T is to a composite number, the greater the reduction in computation. The greatest reduction in computation is achieved when T is an exact power of 2 (e.g., 4, 8, 16, 32, 64, 128, 256, etc.). When this condition is met, the fast Fourier transform may be used. This is why many computer programs for spectral analysis (e.g., SYSTAT, LabVIEW) demand that the number of input data points in the time series be a power of 2.

When the Fourier transform is applied to the autocovariances of the time series, it results in an estimation of the spectral density function (SDF) of that series. The SDF is a collection of variance estimates. More specifically, the SDF portrays how the variance in the series can be broken down into component periodic oscillations of different frequencies. These collections of spectral density estimates are often graphed by computer programs. If the time series contains one sine wave hidden by random noise, then the graph of the SDF plot (variance accounted for plotted by length of sine-cosine period) would reveal a spike at the frequency of that sine wave. This spike would mean that a sine wave at that frequency accounts for a certain amount of variance in the time series.

For example, take the time series given in Figure 9.1. This is a plot of one person's daily moods for three months. The SDF of this data is graphed in Figure 9.6. It can be seen that there is a spike in the SDF at a frequency defined by a seven-day period. This means that, for this person, their daily moods exhibit an oscillating character that cycles from peak to peak every seven days. In other words, there is a weekly cycle in this person's daily moods. As it turns out, this person's moods tend to peak around Saturday and are at their lowest around Wednesday. This is a very common pattern among college undergraduates, and these results tend to contradict the myth of "Blue Monday," since the worst day of the week, in terms of negative moods, turns out to actually be Wednesday for the majority of college undergraduate students (Stone *et al.* 1985).

If an SDF revealed two spikes, then one would conclude that two periodic components were contained in the series. The logic extends to any number of spikes in the SDF. For example, recall that the data in Figure 9.5 were generated by adding three pure sine waves of different frequency (respectively completing 2, 4, and 7 cycles during the observation period) along with some random noise. If we pass this data set through spectral analysis, we get the SDF displayed in Figure 9.7. This

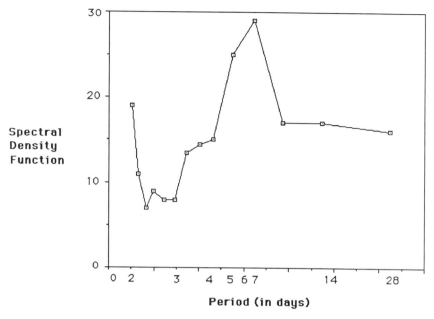

Period (in days)

FIGURE 9.6.   Spectral density function of the data from Figure 9.1 illustrating a seven-day cycle in the daily moods of this subject.

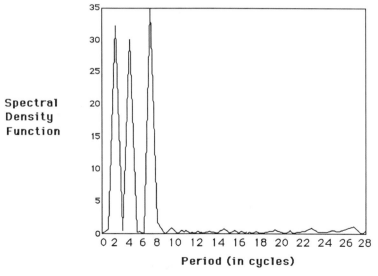

Period (in cycles)

FIGURE 9.7.   Spectral density function of the data from Figure 9.5 illustrating how spectral analysis can recover even hidden periodicities.

figure shows that spectral analysis "found" or recovered the three periodic waveforms that we tried to hide by summing them and adding random noise.

If the SDF revealed a hump instead of a spike, then it would mean that the periodicity was spread over a band of frequencies instead of being concentrated at a single pure frequency. It is in such a manner that spectral analysis is able to reveal hidden periodicities in time-series data. It is useful for exploring and estimating the degree to which some observed process exhibits patterned and repeating changes over time.

3. Assumptions of Spectral Analysis

One assumption of all time-series models is the assumption of stationarity. This assumption refers to the idea that the process generating the observed time series remains stationary and does not change over the period of observation. Strictly speaking, a time series is stationary if the distribution of $X_1, \ldots, X_n$ is the same as the distribution of $X_{1+z}, \ldots, X_{n+z}$. This means that if we shift the time origin by an amount z, it will not have any effect on the distributional characteristics of the time series. This means that it does not matter when in time we begin observing the process, since its distribution does not depend on the value of t. By implication, the covariance between two time points (e.g., the lagged autocorrelations) in a stationary process is a function only of the distance between those points, not a function of the value of t. For the time points t and $t + k$, the value of the covariance between the observations at these time points will be a function only of the lag k, not the starting point of observations in historical time.

In research practice, however, it is useful to define stationarity in a more intuitive and relaxed form. A convenient definition of weak stationarity is that there is no systematic change in either the mean or the variance of the series over time. A systematic change in the mean of a series would denote a trend over time in the data, where the values of the observations drift off in one direction as t increases. A systematic change in variance (e.g., variance changes as t increases) would indicate a process out of control or exhibiting an additive drift, such as occurs in a random walk process where $X_t = X_{t-1} + Z_t$, and Z_t comes from a random number series. In this random walk model the variance increases as t increases so that the series does not conform to a stationary process.

A nonstationary time series can be made stationary in a number of ways. For example, if a trend is apparent in the data, the time series may be detrended. Detrending may be achieved by fitting a least squares regression line to the time series and computing the residuals. These residuals will not contain the trend, but will retain the frequency characteristics of the original series and so will satisfy the "stationarity of mean" assumption. Two other ways to remove trend and lessen the variability of the series is through either smoothing or differencing. Smoothing uses a moving-average transformation on the series

$$X_t = \frac{X_t + X_{t-1}}{2}$$

In this formula two adjacent time points are averaged, and this averaging process moves along the series. Smoothing acts as a "filter" that alters the frequency characteristics of the series in specific ways (see Gottman, 1981). Differencing is achieved by subtracting components of the series. First-order differencing can be defined as

$$X_t = X_t - X_{t-1}$$

Box and Jenkins (1976) provide an extensive discussion of differencing. Note that smoothing and differencing can radically alter the frequency characteristics of the observed data. For example, differencing can introduce periodic variation in a purely random series (Yule, 1981). Thus these transformations should be used carefully.

A third aspect of a stationary series is that it contains no deterministic periodic components. Deterministic periodic components define precise and regular cycles. We can distinguish probabilistic cycles from deterministic cycles, however. A probabilistic cycle occurs *around* a particular frequency, whereas a deterministic cycle has a fixed frequency. This distinction is one of degree rather than dichotomy. And spectral analysis can be used to detect and estimate the frequency of both deterministic and probabilistic periodicities in the data. The point is that, to achieve *strict* stationarity, any deterministic periodic components in the data must be estimated and removed.

A second major assumption of time-series analysis is the assumption of equidistant time points in the observed data. That is, these techniques assume that the time interval between successive observations is constant across the series of measurement occasions. In research practice, this simply means that observations should be regularly spaced in time and missing data should be avoided. Equidistant time points are needed to ensure that the frequency parameters that result from spectral analysis can have a straightforward interpretation in terms of the original time units. Researchers should be aware of how their measurement operations or instruments might violate the equidistant time point assumption. For example, cardiac measures based on the interbeat interval are sequential but are not based on equidistant time points. Thus this physiological measure does not result in observations that are appropriate for spectral analysis. Statisticians are developing complex models for nonequidistant observations (e.g., Bloomfield, 1976; Chatfield, 1980, Chap. 10), but it would take us too far afield from our purposes there to discuss these models.

In terms of missing data points, one might substitute the expected value of a data point for a missing data point (see chapter on missing data estimation in volume one). The expected value of an element in a time series would actually be the mean of the series. Some spectral analysis computer programs will allow for substituting the mean for missing values (e.g., SPSS Version 8.3). Other spectral programs do not allow any missing data (e.g., SAS). Although it is really a judgment call, it seems wise to avoid analyzing a time series with more than 10% of the points estimated by insertion of the mean.

A final logistic consideration of spectral analysis concerns the number of time points that need to be observed. Since estimation is involved, the law of large numbers suggests that the more observations the better, at least up to a point of diminishing returns for the added effort. In practice, most researchers have at least 60 consecutive observations. A related concern involves the observation rate or the length of time between successive observations. The fastest frequency we can detect in a time series (called the *Nyquist frequency*) is one that completes itself every two points of observation. Researchers must be careful to make observations at a rate capable of detecting the fastest frequency they believe the data will exhibit. If the data fluctuate at a faster frequency than the rate of observation, then the problem of aliasing will occur. *Aliasing* simply refers to the erroneous frequency conclusions that may occur when the actual frequency in the observed process is faster than the fastest detectable frequency based on the rate of observation.

4. Spectral Analysis of Psychological Data: A Worked Example

Because spectral analysis detects and quantifies frequency information, it is ideal for use in research where frequency of change needs to be assessed (e.g., Wade, Ellis, and Bohrer, 1973). An example of this is the conceptualization of mood variability as *frequent* changes in mood over time. Many researchers have studied mood variability (Cattell, 1973; Clum and Clum, 1973; Depue et al., 1981; Ekenrode, 1984; Epstein, 1983; Folstein, DePaulo, and Trepp, 1982; Gorman and Wessman, 1974; Johnson and Larson, 1982; Larson, 1983; Larson, Csikzentmihalyi, and Graef, 1980; Linville, 1982; Stallone et al., 1973; Tobacyk, 1981; Wessman and Ricks, 1966), but they index this variability by using a within-subject standard deviation (or variance) of mood scores computed over time for each subject. For example, if mood were measured each day for 75 days, one could compute the standard deviation of such scores over time for each subject. This standard deviation would tell you the average magnitude of fluctuation in mood over the time period observed. But would this index really tap into the *frequency* of mood change over time for each subject?

To answer this question, imagine two subjects depicted in Figure 9.8, on whom we have daily mood reports for 75 consecutive days. Suppose our imaginary subject A exhibits moods as portrayed in Figure 9.8(a) and that imaginary subject B exhibits daily moods as portrayed in Figure 9.8(b). Note that subject A exhibits moods that alternate frequently between positive and negative. Subject B, on the other hand, exhibits a mood profile that goes from extreme positive, through extreme negative and back to extreme positive over the course of the 75-day observational period. Subject B thus exhibits only one change in direction of his or her moods over time. We would *not* say that Subject B has *frequently* changing mood states, even though the average *magnitude* of his moods on any given day might be extremely positive or negative. Subject A, on the other hand, does exhibit *frequently* changing moods.

What would the standard deviation of mood scores tell us about these two subjects? Obviously, the standard deviation index of mood variability would not differentiate between these two subjects at all since both subjects are about the same in terms of their *average magnitude* of change. Even though subjects A and B would obtain very similar standard deviations, it is visually obvious that they are quite different in their patterns of mood variability. The standard deviation contains no information about the serial dependencies in the data, only on the average magnitude of change. The standard deviation itself is actually a type of average score—the mean of the deviate scores. Thus the standard deviation disregards *frequency* of change information in favor of estimating the average *magnitude* of change. Thus when mood variability is

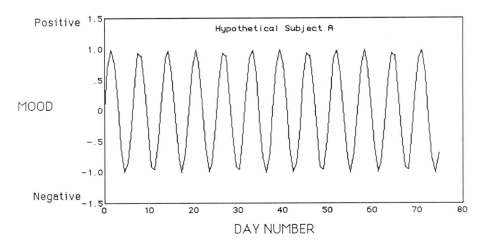

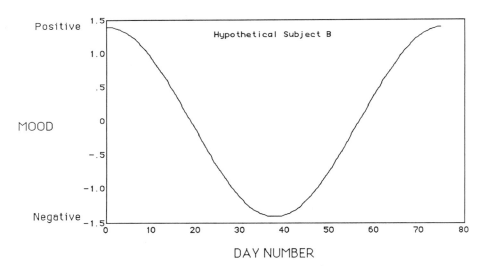

FIGURE 9.8.   Daily mood data for two hypothetical subjects.

conceptualized as the frequency of change, we see that the within-subject standard deviation falls short of providing an index of mood variability.

Spectral analysis, on the other hand, is ideally suited for estimating the frequency parameters of change because it takes into account serial dependencies in the data. Spectral analysis will quantify the degree to which a person's moods are changing rapidly or slowly. Consequently, spectral analysis would be the analytic tool of choice for assessing mood variability as the *frequency* of mood changes over time.

Larsen (1987) applied spectral analysis to the problem of assessing mood variability. The question Larsen's (1987) research addressed was whether a questionnaire measure of emotional reactivity—the Affect Intensity Measure (AIM; Larsen and Diener, 1987)—predicts the frequency of mood change over time. This was a validity study in which mood variability (i.e., *frequency* of mood change over time assessed via spectral analysis) was the criterion to be predicted from the single-session questionnaire measure. In one study reported by Larsen (1987) there were 76 subjects who kept daily records of their moods for 84 consecutive days. Each subject's daily mood data were conceptualized as a single time series containing 84 points (one mood score for each day for three consecutive months).

Because some subjects exhibited slight trends in their daily moods over the three-month observational period, all subjects' daily mood data were detrended using ordinary least squares regression. The residuals from each subjects' daily data were then input into spectral analysis separately for each subject. Fourteen lags were used, and a Bartlett window was applied (Gottman, 1981). Smoothed power spectra were thus estimated for periods ranging from 2 to 28 days in frequency. This means that sine and cosine waves were fitted to each subjects' data ranging from 2 to 28 days in the length of their period. Each spectral estimate represents the amount of variance accounted for by specific sine-cosine wave functions in each subject's data. For illustration, the spectral results from two subjects are presented in Figure 9.9. For subject 24 it can be seen that variability in his data can best be accounted for by fitting very fast sine-cosine waves. Estimates of variance accounted for in this subject's data are relatively large for the shorter periods, illustrating that this person's moods changed frequently over time. Subject 56 is just the opposite; variance in this subject's data is best accounted for by fitting slower sine-cosine waves, meaning that this person's moods changed, for the most part, more slowly over time.

In order to examine the relationship between frequency of mood change and scores on the AIM, it was decided to "chunk" the spectral estimates into what can be called fast, medium, and slow components of mood change speed. Spectral estimates of variance accounted for by the five periodic components of less than three days in length were averaged to yield a composite fast index. A similar slow index was formed by averaging the five spectral estimates from the periodic components

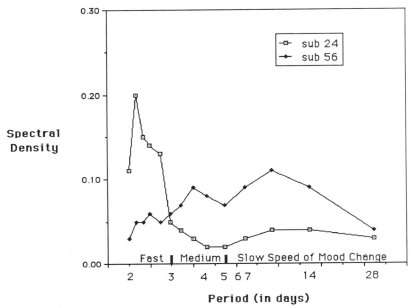

FIGURE 9.9.   Spectral density plots for two subjects (from Larsen, 1987).

ranging from 5.6 to 28 days. And a composite medium index was computed by averaging the four remaining spectral estimates for the periods ranging from approximately three to five days. These divisions for cutting up the estimates into slow, medium, and fast components of mood change speed are illustrated in Figure 9.9.

The AIM was used as the predictor against which to evaluate the spectral approach to assessing mood change frequency because the AIM is believed to assess emotional variability (Larsen and Diener, 1987). Extreme groups on the affect-intensity dimension were created by taking the 25 highest-scoring and the 25 lowest-scoring subjects on the AIM. Plots of the percent of variance sccounted for by summing the fast, medium, and slow spectral estimates are presented in Figure 9.10 for the high and low AIM groups separately. It can be seen that the variance in the daily moods of high AIM subjects is best accounted for by the faster sine-cosine waves, suggesting that their moods changed more frequently relative to the low AIM subjects. For the low AIM subjects, just the reverse was found; variability in their daily moods was best accounted for by fitting the slower sine-cosine waves, indicating that their moods changed less frequently compared with the high AIM subjects. A 2×3

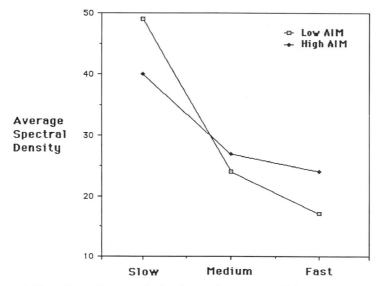

FIGURE 9.10. Plot of spectral density estimates for high- versus low-affect intensity groups.

mixed-design analysis of variance supports what is visually obvious in Figure 9.10. The AIM group (high versus low) was the between-subject factor, and mood change frequency (fast, medium, slow) was the within-subject repeated-measures factor. The average spectral estimate at each level of mood change frequency was the dependent variable. No significant effect was found for the AIM group. The effect for level of mood change speed was significant $[F(2, 48) = 150.9, p < .091]$, indicating that more variance was associated with slower changes in mood than with fast changes in mood. Of more interest, however, was the interaction between the AIM group and the speed of mood change, which proved to be significant $[F(2, 48) = 13.3, p < .001]$. This interaction implies that for the high-AIM group, variability in daily moods occurs with more rapid or more frequent shifts, whereas the low-AIM subjects tended to exhibit moods that change more slowly over time.

Another question that can be addressed using spectal analysis of these data is the issue of sex differences in the frequency component of mood change. For example, if menstrual cycle has its purported effect on mood, then women should show more variability in their daily moods associated with a monthly periodic component. That is, women and men should *not*

TABLE 9.1. Sex differences in Spectral Density Estimates for Periodic Components of Various Lengths in Daily Moods

Period length (in days)	M (men)	M (women)	t	Significance of difference
2.00	.064	.065	.005	ns
2.15	.062	.063	.011	ns
2.33	.063	.062	−0.07	ns
2.55	.058	.058	0.03	ns
2.80	.058	.060	0.54	ns
3.11	.061	.070	1.64	ns
3.50	.069	.082	2.03	ns
4.00	.072	.074	0.32	ns
4.66	.077	.077	−0.04	ns
5.60	.105	.104	−0.09	ns
7.00	.144	.133	−1.09	ns
9.33	.141	.125	−2.01	ns
14.00	.148	.151	0.16	ns
28.00	.137	.185	2.26	.02

differ in any of the spectral estimates *except* for that spectral estimate associated with a 28-day periodic component.

Spectral analysis is well suited to address this question because it is phase-independent. It can identify cyclical components in a time series regardless of what phase that cycle is in when observation of the process begins. Consequently, if there is a monthly cycle for mood in women, then it makes no difference what phase of that cycle they were in when the study begain, and it does not matter if the women were each in a different phase of their cycle when mood reporting started.

Table 9.1 presents the average spectral estimates of mood change frequency for each periodic component from 2 to 28 days for the men and women in this sample. t-Tests were used to assess the significance of the differences between men and women at each period length. It can be seen that the only significant differences occurs for mood variance associated with an approximately 28-day cycle, with women showing more 28-day cyclical variability in thier moods than men. It might be argued that, because 14 t-tests were conducted, at least one of them should reach significance by chance alone. However, each t-test has an equal probability of obtaining significance by chance. Therefore the probability that the single test predicted to reach significance actually *did* reach significance in the predicted direction is actually quite small.

These findings suggest that women have significantly more variability in their daily moods associated with a monthly cycle than do men. These results do not, however, mean that this periodic component of mood change is associated with menstrual cycle. Even though this seems to be a likely inference, it is unwarranted because phase of menstrual cycle was never assessed by the female subjects in this sample. We do not know, for example, if negative mood leads, lags, or coincides with menstrual cycle in women. Nevertheless, because spectral analysis is useful in identifying individual and group differences in components of mood change frequency, it may prove extremely useful for research on such topics.

In these examples each subject's data are viewed as a single time series for input into spectral analysis. Using this approach, we can examine the parameters of mood change frequency over time for single subjects. Spectral analysis estimates how much variance in the time series can be accounted for by the periodic components of various lengths. Then the frequency parameters from spectral analysis can themselves be examined for meaningful individual or group differences. This strategy is well suited for the study of *any* phenomenon that varies over time and where the research question concerns possible periodic components that might underlie that variability. The application of spectral analysis to psychological questions is limited only by the creativity and ingenuity of the researcher willing to collect time-series data.

5. Bivariate Spectral Analysis

Spectral analysis is a method for estimating the frequency components of a series of measurements made on single cases over time. What if the researcher has measured two variables on each occasion over time and wants to assess the degree of rhythmic co-occurrence or temporal covariation between these variables for each subject? For this problem cross-spectral analysis offers a solution.

Cross-spectral analysis is, in general terms, used to estimate the degree of rhythmic coherence between two separate time series. Specifically, just as univariate spectral analysis is the Fourier transform of the auto-covariance function, cross-spectral analysis is the Fourier transform of the cross-covariance function. Consqeuently, cross-spectral analysis estimates the degree of association between the rhythmic components of two separate time-series at specific frequencies. The output of cross-spectral analysis is a collection of coherence estimates, each one revealing the proportion of variance accounted for by the influence of one series on the other at each specific frequency.

Porges et al. (1980) provide an excellent example for illustrating cross-spectral analysis. Their example concerns the degree of rhythmic co-occurrence between heart rate and respiration. These two physiological systems are inherently rhythmic and somewhat related to each other in a complex manner. The purpose of their example is to assess the degree of rhythmic heart rate variance that is associated with respiration. The computation of cross-spectral coherence estimates between heart rate and respiration is demonstrated. Implicitly this is done for each subject separately, and elsewhere (Porges, 1983) it is demonstrated that the coherence of heart rate and respiration is useful in assessing central nervous system dysfunction in infants and other aspects of fetal distress. This example illustrates how cross-spectral analysis can be used to assess the degree to which two phenomenon are "coupled" over time and how this "coupling" can be useful in understanding other individual or group differences (e.g., fetal distress).

The application of cross-spectral methods is not limited to physiological measures. Certainly many physiological processes are inherently rhythmic and so spectral methods appear to be the tools of choice to quantify such periodicities. Nevertheless, researchers may be interested in behavioral regularities and the rhythmic co-occurrence between behavioral systems over time. Or perhaps researchers might be interested in the co-occurrence between biological and behavioral systems within individuals. In such instances cross-spectral methods will be useful tools for quantifying the degree of shared rhythmic variation at specific frequencies for each case. Then these estimates may be used to test hypotheses regarding individual or group differences in the degree of within-subject coherence between the variables of interest.

Porges et al. (1980) also present an important quantitative extension of the cross-spectral coherence estimate that may prove useful to researchers in psychology. A limitation of the traditional cross-spectral coherence analysis is that it gives estimates of shared rhythmic variance only at *specific* frequencies. Researchers may, however, have a need for a summary statistic that can describe the proportion of shared variance between two time series over a *band* of frequencies. Porges et al. (1980)

develop the weighted coherence estimate to describe the general degree of coupling that exists between the rhythmic properties of two time series over a band of frequencies. This is important to social scientists because rhythmicities in the phenomena we study are usually, if at all, manifest over a range of frequencies; *pure* sinusoidal rhythms are quite rare in living systems, especially when the phenomenon under study is multiply determined. Traditional cross-spectral coherence estimates, because they focus only on specific frequencies, work best for relatively pure rhythmic processes, such as pure tone sound waves or electrical energy. Biological and social systems, on the other hand, may produce *roughly* periodic processes that are contaminated with either random error, various degress of disruption, or overlaid with other periodic effects (i.e., general stochastic processes). Because such processes would exhibit one or more *bands* of dominant frequencies, the weighted coherence estimate is a better statistic for describing the temporal coupling of two such processes than the traditional coherence estimates.

6. Computer Programs for Spectral Analysis

6.1. Mainframe Packages

The Statistical Analysis System (SAS) package for mainframe computers has an extensive collection of time-series analysis procedures. These are collected together in an optional package called SAS/ETS, for econometric time series. Within this collection there are extensive capabilities for handling time-series data as input and controlling output. More specifically, there is a procedure for spectral analysis that is quite flexible. It can handle both univariate and bivariate time series and will do an optimal test for whether the series is white noise. The spectral procedure allows for the smoothing and/or zero-centering (removing the mean) of input data. A convenient aspect of mainframe packages such as SAS is that the data may be detrended in the regression module and the residuals sent directly to the spectral analysis module. The spectral procedure in SAS will abort and issue an error message if it encounters a missing data point; there are no options for substituting the mean for missing values within the spectral procedure. If T (the number of input values) can be factored into a prime integer less than 24, then the fast Fourier transform is used, otherwise a much slower method is used. Output is highly user-controlled. For example, it is possible to express frequencies in cycles per observation unit. One distracting feature of SAS spectral analysis is that the spectral density values must be plotted by calling a separate plot procedure and inputting the values of the periodogram that were output from the spectral procedure.

The Statistical Package for Social Sciences (versions SPSS and SPSSX) has no formal spectral analysis procedures in their base products. However, SPSS-6000 Release 8.3, prepared by the Northwestern University Vogelback Computing Center, does contain a spectral analysis routine. This release is fairly common among installations that have older mainframe computers from Control Data Corporation (CDC). The spectral analysis procedure in this release of SPSS is also quite flexible. It can accept univariate or bivariate input, it allows for the substitution of means for missing data points, and it allows the user to select from a variety of standard windowing options. The output is also very much under the user's control, with options for plotting autocovariance or autocorrelation functions (weighted or unweighted), the SDF, the log of the SDF, and various indices such as the bandwidth of the chosen window, degrees of freedom, and total variance accounted for. Output for bivariate spectral analysis is also quite extensive, including coherence estimates, cross-covariance functions, cospectrums, and gain and phase-shift parameters. A convenient aspect of this procedure is that it plots various parameters without having to call other plotting routines.

The new version of SPSS-X (Release 3.0) offers an optional enhancement called Trends. Trends includes many facilities for handling and analyzing time series data, including a very flexible spectral analysis procedure. The Trends package itself provides many procedures for manipulating time series variables, such as the replacement of missing values with series mean or other values interpolated from the whole series or local segments of the series. It also provides for on-line smoothing, differencing, and detrending by linear or nonlinear regression on the same run as the formal time series analysis. Trends also contains a very flexible plotting procedure that can output to a PC graphics program or to SPSS's own graphics facility. The spectral analysis procedure provides both univariate and bivariate analyses, can apply a variety of spectral windows, plots most of the variables generated by the analysis (e.g., the periodogram, spectral density estimates, coherence estimates, etc.), and can save generated variables or command language for later runs. Although Trends was only recently released, it appears to be a quite flexible and powerful program for spectral analysis that will most likely be available at installations that provide the SPSS-X Release 3.0 base package.

Another package contained on most mainframe computers is the International Mathematical and Statistical Library (IMSL). This package represents a collection of FORTRAN-callable subroutines for specific kinds of numeric processing. There are several attractive features of the IMSL package. For example, it is quite flexible since the user constructs the step-by-step processing of the data with a FORTRAN program. IMSL is also implemented on most university-based supercomputer installations. At some sites, many of the IMSL subroutines have been vectorized to take full advantage of the supercomputer's optimal processing speed. IMSL contains spectral and Fourier transform routines for both univariate and bivariate time-series data.

6.2. Personal Computer Statistical Packages

Both SAS and SPSS have versions for use on personal computers (PC), but neither of these PC products contain a spectral analysis procedure. One fairly common PC statistical package, however, does contain a spectral analysis module. SYSTAT is a relatively complete statistical package that is available for both IBM-compatible PCs as well as Apple Macintosh computers. SYSTAT contains a time-series module that allows for a good deal of data preparation (e.g., smoothing) as well as a variety of time-series analyses. For spectral analysis with this package one must use a Fourier decomposition routine with the constraint that the number of data points must be a power of 2. Then the real component of the Fourier series is squared and plotted against the frequency index. To window the periodogram, it is necessary to smooth the output of the Fourier analysis before plotting it.

Many PC statistical programs that are designed for data acquisition contain spectral analysis routines. Such programs are designed for the transformation of analog input signals into digital values for analysis. These data acquisition programs are often used in electrical engineering applications and in psychophysiological applications, where the data are time-based electrical signals emanating from some source (e.g., a human or animal subject or an electrical system). Once the data are digitized they are stored for later analysis by these programs, including spectral decomposition programs. The point is that the data do not *have* to come through some analog input device in order to use the spectral analysis program. In fact, one can input any data file directly into such packages for analysis. One such package, available for IBM-compatible PCs and Apple Macintosh computers, is LabVIEW from National Instruments. This package contains a wide variety of data conditioning and analysis routines. In fact, such programs have more flexibility for data conditioning than most statistics-only packages because they have been designed for the analysis of complex electrical signals. The disadvantage to such programs is the engineering jargon that must be digested in the documentation for the programs. For example, instead of being referred to as a time series, the input data is often called a *signal array,* and the various conditioning and analysis routines are often called *digital signal processing algorithms.* Electrical engineers make wide use of spectral analysis for describing the frequency properties of electrical signals. Nevertheless, despite the electronics jargon, these packages actually do a fine job of traditional spectral analysis. In LabVIEW, the input time series is restricted to 1024 time points with the additional restriction that the number of time points must be a factor of 2, since only the fast Fourier transform is implemented. Both univariate autocorrelation and bivariate cross-correlation functions are available, as well as spectral and cross-spectral analysis routines. Convolution and deconvolution routines are available, as well as real and inverse fast Fourier transform routines. A variety of discrete integration and differentiation functions are available, along with input reversing, clipping, function shifting, and pulse pattern analysis. Four different types of common windows are built into LabVIEW, although the user may construct any type of specialized window or filter using a mathematical function routine. Finally, LabVIEW has several signal or data generation routines, where the user may create and analyze pure or noisy sine waves, pulse or impulse waves, ramp or triangle waves, or white or Gaussian noise waves. In fact, Figures 9.3–9.5 were generated using LabVIEW on a Macintosh computer. Experimenting with the spectral analysis of self-generated data, where the properties of the input data are known, is especially instructive to the person just starting out with spectral analysis. This is possible with any computer software that allows for generating data with different frequency properties.

Finally, there exist a number of stand-alone fast Fourier transform or spectral analysis programs available for personal computers. One that I am familiar with is contained in the Reed Applications II disk developed for Macintosh computers by the Reed Development Laboratory at Reed

College and distributed by Kinko's Academic Courseware Exchange. Two programs are available on this disk. One program (FFT) takes a time-series data set from a file or from the keyboard and computes its spectrum using the fast Fourier transform routine. This program will display the output (i.e., the spectral density estimates) as either text or graphics and will optionally store the output to disk. The other program (Signal) allows the user to draw an input signal using the Macintosh mouse and then it analyzes that signal using the fast Fourier transform and automatically displays the spectral density function in a graphics window next to the input signal. Several built-in data sets are available in this program, including sine, square, pulse, noise, damped sine, AM, FM, and burst waves. This program seems more useful as a heuristic learning aid than as an actual data-analytic tool, because it shows what the spectral density function would look like for input time series of any conceivable shape or pattern.

References

Box, G. E. and Jenkins, G. M. (1976). *Time-series analysis, forecasting and control,* San Francisco: Holden Day.

Bloomfield, P. (1976). *Fourier analysis of time-series: an introduction.* New York: Wiley.

Brillinger, D. R. (1975). *Time-series data analysis and theory.* Chicago: Holt, Rinehart and Winston.

Cattell, R. B. (1973). *Personality and mood by questionnaire.* San Francisco: Jossey-Bass.

Chatfield, C. (1980). *The analysis of time-series: theory and practice,* London: Chapman and Hall.

Clum, G. A., and Clum, J. (1973). "Mood variability and defense mechanism preference," *Psychological Reports.* **32,** 910.

Depue, R. A., Slater, J. F., Wolfstetter-Kausch, H., Klein, D., Goplerud, E., and Farr, D. (1981). "A behavioral paradigm for identifying persons at risk for bipolar depressive disorder: a conceptual framework and five validational studies," *Journal of Abnormal Psychology Monograph.* **96,** 381–437.

Eckenrode, J. (1984). "Impact of chronic and acute stressors on daily reports of mood," *Journal of Personality and Social Psychology.* **46,** 907–918.

Epstein, S. (1983). "A research paradigm for the study of personality and emotion." In M. M. Page (Ed.), *Personality—current theory and research: 1982 Nebraska symposium on motivation* (pp. 91–154). Homewood, Ill.: Dorsey Press.

Folstein, M. F., DePaulo, J. R., and Trepp, K. (1982). "Unusual mood stability in patients taking lithium," *British Journal of Psychiatry*. **140**, 188–191.

Gorman, B. S., and Wessman, A. E. (1974). "The relationship of cognitive styles and moods," *Journal of Clinical Psychology*, **30**, 18–25.

Gottman, J. M. (1981). *Time-series analysis: a comprehensive introduction for social scientists*. Cambridge, England: Cambridge University Press.

Johnson, C., and Larson, R. (1982). "Bulimia: An analysis of moods and behavior," *Psychosomatic Medicine*. **44**, 341–351.

Larsen, R. J. (1987). "The stability of mood variability: A spectral analytic approach to daily mood assessments," *Journal of Personality and Social Psychology*. **52**, 1195–1204.

Larsen, R. J., and Diener, E. (1987). "Affect intensity as an individual difference characteristic: a review," *Journal of Research in Personality*. **21**, 1–39.

Larson, R. W. (1983). "Adolescents' daily experience with family and friends: contrasting opportunity systems," *Journal of Marriage and the Family*. **45**, 739–750.

Larson, R. W., Csikzentmihalyi, M., and Graef, R. (1980). "Mood variability and the psychosocial adjustment of adolescents," *Journal of Youth and Adolescence*. **9**, 469–489.

Linville, P. W. (1982). "Affective consequences of complexity regarding the self and others." In M. Clark and S. Fiske (Eds.), *Affect and cognition: seventeenth annual symposium on cognition* (pp. 79–109). Hillsdale, N. J.: Lawrence Erlbaum.

Porges, S. W. (1983). "Heart rate patterns in neonates: A potential diagnostic window to the brain." In T. Field and A. Sostek (Eds.), *Infants born at risk* (pp. 3–22). New York: Grune and Stratton.

Porges, S. W., Bohrer, R. E., Cheung, M. N., Drasgow, F., McCabe, P. M., and Keren, G. (1980). "New time-series statistic for detecting rhythmic co-occurrence in the frequency domain: the weighted coherence and its application to psychophysiological research," *Psychological Bulletin*. **88**, 580–587.

Stallone, F., Huba, G. J., Lawlor, W. G., and Fieve, R. R. (1973). "Longitudinal studies of diurnal variation in depression: a sample of 643 patient days," *British Journal of Psychiatry*. **123**, 311–318.

Stone, A. A., Hedges, S. M., Neale, J. M., and Satin, M. S. (1985). "Prospective and cross-sectional mood reports offer no evidence of a "blue monday" phenomenon," *Journal of Personality and Social Psychology*, **49**, 129–134.

Tobacyk, J. (1981). "Personality differentiation, effectiveness of personality integration, and mood in female college students," *Journal of Personality and Social Psychology*. **41**, 348–356.

Wade, M. G., Ellis, M., and Bohrer, R. (1973). "Biorhythms in the activity of children during free play activity," *Journal of Experimental Analysis of Behavior*. **20**, 155–162.

Wessman, A. E., and Ricks, D. F. (1966). *Mood and personality.* New York: Holt, Rinehart and Winston.

Yule, G. U. (1981). "A method of investigating periodicities in disturbed series, with special reference to Wolfer sunspot numbers." In A. Stuart and M. Kendall (Eds.), *Statistical papers of George Udny Yule.* New York: Hafner Press.

Univariate and Multivariate Time-Series Models: The Analysis of Intraindividual Variability and Intraindividual Relationships*

Chapter 10

BERNHARD SCHMITZ

*Max-Planck Institute
for Human Development and Education
Berlin, Federal Republic of Germany*

Abstract

The tasks of developmental psychology include the analyses of intra-individual variability and intraindividual relationships. This article describes methods appropriate for these tasks: univariate and multivariate ARIMA models. The latter are of particular importance because they allow the investigation of time-lagged relationships, which can be interpreted as "causal" relations. It is argued that multivariate ARIMA models are preferred to the analysis of cross-correlations. Applications of these models are presented for simulated time series as well as for empirical data.

1. Introduction

Before beginning a search for methods appropriate to developmental psychology, one must first focus on the proposed tasks of the discipline. As formulated by Baltes, Reese, and Nesselroade (1977), the tasks of developmental psychology are to describe, explain, and modify (optimize) intraindividual change in behavior and interindividual differences

* The author thanks Bettina Törpel for assistance during the preparation of this chapter.

351

in such change across the life-span. Since elaborated theories about the course of development (such as the compensation hypothesis; see Baltes and Willis, 1982) are generally not confined to one variable but rather include assumptions about interactions between several variables, another task to those already cited should be added: to examine intraindividual relationships.

To analyze intraindividual change, the same variable must, at minimum, be measured on the same individual at two points in time. However, serious problems arise when change is measured in this simple way. It appears more advisable in many cases to include measurements taken at far more than just two points in time. A number of such measurements, taken from a single observational unit, constitutes a time series. A broader definition of time series appropriate to psychological data is given by Gregson (1983): "Psychological data are segments of life histories; as such they are ordered sequences of observations and by definition time series" (p. ix). There are several arguments for the collection of time-series data in order to study psychological processes and to analyze these data with time-series methods.

1. *To avoid theoretical problems concerning measurement.* The initial premise of classical test theory is that a person's "true score" is constant over time and that therefore all deviations from the true score are caused solely by error of measurement. Accordingly, classical testing runs into difficulties when deviations, or changes, are measured and interpreted as meaningful. With measurements taken at only two points in time, there also arises the problem of regression to the mean, or, more generally, the problem that later values depend on the initial one (see Harris, 1963). Time-series analyses avoid these difficulties.

2. *To determine the shape of the developmental curve.* With time-series data, developmental trajectories may be tested. For example, a simple hypothesis might be a phase model (see Kohlberg, 1963; Piaget, 1948). The testing of even such a simple model demands a number of measurements if the precise transitions from one phase to the next are to be ascertained. An even greater number of observations would be needed for hypothesized nonmonotonic trajectories.

3. *To ascertain intraindividual variability.* The range of intraindividual variability deserves careful examination. Estimating *intra*individual variability on the basis of *inter*individual variability, as done in classical test theory, is a dubious approach because the two forms of variability can differ substantively.

4. *To test for intervention effects.* Time series allow a statistical

assessment of intervention effects for individual cases. The confirmation of a hypothesis regarding the effectiveness of an intervention represents an explanation of intraindividual variability. Statements about the onset and duration of the effect can be made as well, something not possible with measurements taken at only two points in time. Ultimately, inferences about the optimization of intervention can be drawn from such analyses.

5. *To facilitate the analysis of intraindividual relationships.* The observation of intraindividual relationships requires analyses of multivariate time series.

6. *To observe dynamic (lagged) intraindividual relationships and "causal" mechanisms.* In addition to synchronic relationships between variables, time series permit the study of asynchronic relations, giving clues to causality and thereby also helping to further explain intraindividual variability.

7. *To avoid statistical problems in the analysis of repeated measurements.* In the analysis of time-series data, one can avoid such statistical problems as the serial dependence of data. Customary analyses of variance results are skewed by such dependence.

Arguments 1 through 6 focus on the analysis of individual cases. Psychologists generally wish to extend their knowledge beyond single individuals, and therefore a combined use of both time-series and cross-sectional analyses may be justified. In this case, time-series data would be recorded for each person in a sample of individuals. Data would be statistically analyzed for each individual separately, and resulting time-series parameters for each person would constitute the data base for cross-sectional analyses. Such a design may represent a solution to the long-standing controversy (see Allport, 1962) between idiographics and nomothetics. Such a procedure offers several advantages. First, two or more individuals' parameters become statistically comparable. Second, the averaging of individual trajectories, which can lead to a functional type that does no justice to any individual trajectory, becomes unnecessary. Instead, similar trajectories can be combined. Third, the necessary distinction between intraindividual and interindividual relationships becomes possible. Customarily, analyses are made interindividually (cross-sectionally), and therefore do not necessarily allow valid inferences about intraindividual relationships.

Although the advantages of this combined design have been briefly detailed here, lack of space makes it necessary to confine this introductory section to a discussion of the analysis of individual cases.

The following presents time-series procedures in relation to the aforementioned tasks of research in developmental psychology. The emphasis is on the multivariate analysis of dynamic relations between time series, a hitherto little used procedure. The presentation provides a general introduction and highlights new developments in the field (e.g., the extended sample autocorrelation function and model selection criteria not included in standard works (Glass, Willson, and Gottman, 1975; Gottman, 1981; McCleary and Hay, 1980).

In Section 1, procedures for describing time-series data will be briefly presented. Section 2 (clarification of basic terms), Section 3 (description of time-series models), and Section 4 (the process of analyzing time series) all deal with univariate analyses and prepare the reader for an understanding of the multivariate analyses presented in Section 5.

1.1. Description of Time-Series Data

Time-series data should be graphically represented before being statistically analyzed. The mean and variance of a time series, determined in the same way as for a cross-sectional analysis, should also be obtained. Linear, quadratic, or higher-order polynomial trends can be fitted to the time series for a description of the long-term trajectory. Spectral analytical procedures also lend themselves to the description of such periodic trajectories as weekly, monthly, or yearly rhythms (see Gottman, 1981; Larson, this volume).

2. Basic Terms

2.1. The Lag Function and the Backshift Operator

For time-series analysis, the comparison between current measurement and measurement taken at some previous point in time is often important. An example would be the comparison of an original series of daily measurements with the series from the previous day. If z_t is the original series of daily measurements, the series from the previous day, a series of lag 1, is expressed as z_{t-1}. In general, the series of lag k is expressed as z_{t-k}. The *backshift operator B* denotes the procedure by which a time series is shifted back by one lag. Applied to z_t, the backshift operator provides the lagged time series z_{t-1}. The corresponding equation is $B(z_t) = z_{t-1}$. Applying B to z_{t-1}, one gets z_{t-2}, or $B(z_{t-1}) = z_{t-2}$.

One can also write: $B(B(z_t)) = z_{t-2}$, or $B^2(z_t) = z_{t-2}$. One can calculate with the backshift operator as with mathematical symbols.

2.2. Autocorrelation

If T is defined as the number of observations, then the correlation of a time series with the same series, lagged, is called an *autocorrelation:* $r_1 = r(x_t, x_{t-1})$ or, more generally,

$$r_k = r(x_t, x_{t-k}) = \frac{\sum_{t=k+1}^{T} (x_{t-k} - \bar{x})(x_t - \bar{x})}{\sum_{t=1}^{T} (x_t - \bar{x})^2}$$

If one calculates autocorrelations for a series with the same series of lag $1, 2, 3, \ldots$, this set of autocorrelations can be represented as a function of the lag. Such an *autocorrelation function* (ACF) describes the temporal pattern of serial dependence.

2.3. Partial Autocorrelation

It is possible that the relationship between the original series and the series of lag k is mediated via an intervening point in time (see Figure 10.1). As with partial correlation in conventional statistics, the partial autocorrelation between an original series and the series of lag 2, for example, is the correlation between z_t and z_{t-2}, controlled for z_{t-1}. If one formulates a regression equation

$$z_t = \phi_1 z_{t-1} + \cdots + \phi_k z_{t-k}(+a_t)$$

then the regression coefficient, corresponding to z_{t-k}, ϕ_k, gives the partial autocorrelation to lag k. The series of partial autocorrelations as a function of the lag is called the *partial autocorrelation function* (PACF).

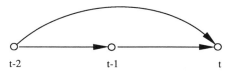

$t-2$ $\qquad\qquad$ $t-1$ $\qquad\qquad$ t

FIGURE 10.1. Direct (upper arrow) and indirect (lower arrow) paths relating time $t - 2$ to t.

2.4. The Stochastic Process as a Time-Series Model

In conventional statistics, the intent is to make statements about one or more populations. In time-series analysis it is confusing to speak of populations because questions of interest primarily concern individual cases. Instead of "population," one therefore frequently refers to "stochastic process." A stochastic process is defined as a sequence of random variables z_t, $t \in \mathbb{N}_0$.

An especially important stochastic process is white noise. White noise a_t is characterized by the independence of a_t and a_{t-k} for all k's.

2.5. Stationarity

In principle, the mean, variance, and autocovariance of a process can be functions of time (t). Processes in which this is the case are called *nonstationary*. Processes whose expectation (mean value) is constant over time are called *mean stationary*. *Covariance* stationarity exists when the autocovariances (and, hence, autocorrelations) are a function of lag but not of time. A process may be assumed to be at least weakly stationary when it is mean and covariance stationary.

3. Univariate ARIMA Models

In psychological applications of time-series analysis one kind of model has become particularly important: the autoregressive integrated moving-average (ARIMA) model. This type of model is composed of three elements: the autoregressive component, the integrated component, and the moving-average component.

3.1. White Noise

White noise a_t is part of any time-series model. It is defined by

$$E(a_t) = 0,$$

$$\text{var}(a_t) = \sigma_a^2,$$

$$\gamma_k = E(a_t a_{t+k}) = \begin{cases} \sigma_a^2, & k = 0, \\ 0, & k \neq 0, \end{cases}$$

$$\rho_k = \frac{\gamma_k}{\sigma_a^2} = \begin{cases} 1, & k = 0, \\ 0, & k \neq 0. \end{cases}$$

Because of the independence of the random variable a_t, the autocorrelations and partial autocorrelations for any lag are 0.

3.2. AR(p) Models

A first-order autoregressive (AR(1)) process is characterized by the equation:[1]

$$z_t = \phi_1 z_{t-1} + a_t.$$

The equation for a general autoregression AR(p) is:

$$z_t = \phi_1 z_{t-1} + \phi_2 z_{t-2} + \cdots + \phi_p z_{t-p} + a_t.$$

The AR equation is an ordinary multiple regression equation, where the preceding values are predictors and the current value is the criterion. This relationship between predictors and criterion accounts for the term *autoregression*. The number of predictors determines the order of the process. The regression weights ϕ_k correspond to a regression's beta weights. To satisfy stationarity conditions in AR processes, order-dependent restrictions for the ϕ_k parameters must be introduced. For instance, for an AR(1) process these restrictions require that $|\phi_1| < 1$.

The AR autocorrelation function decreases exponentially or sinusoidally. Partial autocorrelations corresponding to a lag $> p$ equal 0 (see Figure 10.2, $p = 2$).

3.3. MA(q) Models

A first-order moving-average process (MA(1)) is characterized by the equation

$$z_t = a_t - \theta_1 a_{t-1}.$$

The equation for a general MA(q) process is

$$z_t = a_t - \theta_1 a_{t-1} - \theta_2 a_{t-2} - \cdots - \theta_q a_{t-q}.$$

In conventional statistics the moving average is defined as the weighted average of a value and the adjacent previous and following values. Similarly, an MA(q) process is the weighted average of the current and the q preceding random shocks. The number of previous random shocks contained in the equation determines the order of the process. To ensure

---

[1] z_t are deviations from the mean.

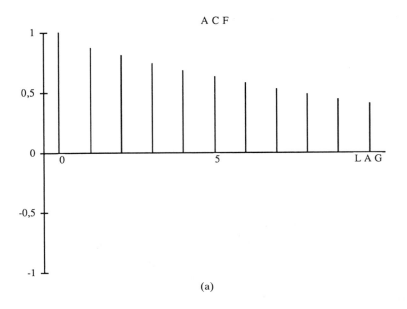

(a)

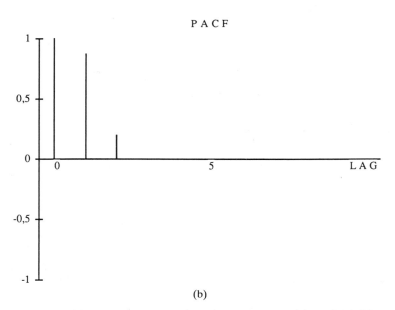

(b)

FIGURE 10.2.   (a) Autocorrelation (ACF) for an AR(2)-model. (b) Partial autocorrelation functions (PACF) for an AR(2)-model.

an unambiguous relationship between parameter and autocorrelation, invertibility conditions must be introduced. For an MA(1) process, the invertibility condition is $|\theta_1| < 1$.

Autocorrelations of MA(q) processes for lags $> q$ equal 0, and the PACF values decrease exponentially or sinusoidally (see Figure 10.3, $q = 2$).

3.4. ARMA(p, q) Models

The system underlying an autoregressive moving-average [ARMA(p, q)] process "stores" p preceding system states and q previous random shocks. Thus an ARMA(1, 1) process is presented as

$$z_t - \phi_1 z_{t-1} = a_t - \theta_1 a_{t-1},$$
$$z_t = \phi_1 z_{t-1} + a_t - \theta_1 a_{t-1}.$$

The general ARMA(p, q) formula is

$$z_t = \phi_1 z_{t-1} + \phi_2 z_{t-2} + \cdots + \phi_p z_{t-p}$$
$$+ a_t - \theta_1 a_{t-1} - \theta_2 a_{t-2} - \cdots - \theta_q a_{t-q}.$$

In empirical investigations ARMA(p, q) time series with $p + q > 2$ rarely occur.

3.5. ARIMA(p, d, q) Models

ARMA processes are only defined for stationary time series. In the case of a nonstationary time series, calculating differences of subsequent process values $z_t - z_{t-1}$ can lead to a stationary ARMA(p, q) model. If stationarity is not achieved after taking the first difference, the differencing operation has to be repeated. The degree of differencing d provides the third component of an ARIMA(p, d, q) model.

3.6. A General Representation of ARIMA Models

The backshift operator allows the introduction of formal notation for the ARIMA models. With

$$Bz_t = z_{t-1},$$
$$B^p z_t = z_{t-p},$$

the AR(p) equation may be written

$$\phi_p(B)z_t = a_t,$$

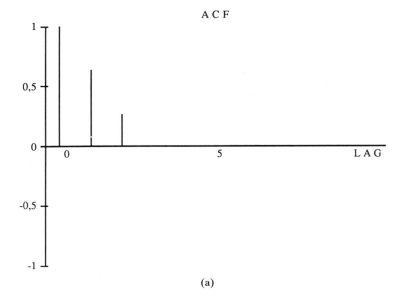

(a)

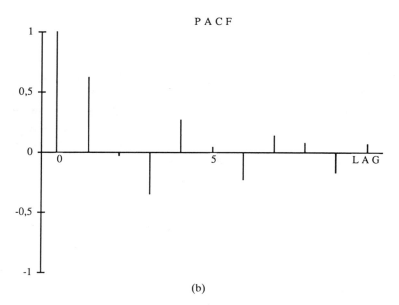

(b)

FIGURE 10.3. (a) Autocorrelation (ACF) for an MA(2)-model. (b) Partial autocorrelation functions (PACF) for an MA(2)-model.

whereas the polynomial is defined as

$$\phi_p(B) = 1 - \phi_1 B^1 - \cdots - \phi_p B^p.$$

An analogous formulation possible for the MA process is

$$z_t = \theta_q(B) a_t,$$

where

$$\theta_q(B) = (1 - \theta_1 B^1 - \cdots - \theta_q B^q).$$

Any differences taken appear in the following notation:

$$(1 - B)^d z_t$$

Thus, it is possible to depict the ARIMA models as

$$w_t - \phi_1 w_{t-1} - \cdots - \phi_p w_{t-p} = a_t - \theta_1 a_{t-1} - \cdots - \theta_q a_{t-q},$$

where

$$w_t = \begin{cases} (1 - B)^d z_t, & d > 0, \\ z_t, & d = 0. \end{cases}$$

Any AR(p) process may be transformed into an MA(∞), and any MA(q) may be transformed into an AR(∞).

4. Conducting Time-Series Analysis

It has been shown that ARIMA models can be characterized by their ACF and their PACF. Therefore, to identify the type and order of an empirical time series, one first calculates the ACF and the PACF and then compares obtained patterns with the same functions of known models. If the pattern of estimates from an empirical time series matches that from a theoretical model, then the identification of the ARIMA model type and order has been successful.

After identification, the next step is to estimate the empirical model's parameters from sample statistics. Lastly, the model is checked diagnostically to establish whether it is indeed appropriate. Should diagnosis show that the identified model is not appropriate, the procedures of identification, estimation, and diagnosis must be repeated.

4.1. Identification

The first step toward identifying a time series should always be to graph data to check for outliers, trends, phases, and periodicities. However, some ARIMA models are very difficult to differentiate graphically. Therefore, it is sometimes easier to discriminate between models using ACF and PACF. It is useful to have a test that checks autocorrelations for significance. An approximation of the standard error (S.E.) of an autocorrelation is

$$\text{S.E.}(r_k) = \left(1 + 2 \sum_{m=1}^{q} r_m^2\right)^{1/2} \Big/ \sqrt{T}, \qquad \text{for } k > q.$$

The quotient of the autocorrelation and its standard error approximates the normal distribution, and therefore $z = r_k/\text{S.E.}(r_k)$ may be used as a simple z-test, with values greater than 2.00 indicating significance. If a large number of tests is involved, it must be taken into account that a few significant results are to be expected at random. Therefore, to check a large number of autocorrelations, one might use a test suggested by Ljung and Box (1978). Here, an entire series of autocorrelations are checked for significance as a whole.

4.1.1. The Ljung-Box Q-Test. This *portmanteau test* builds on the statistic

$$Q(k) = T(T + 2) \sum_{m=1}^{k} (T - m)^{-1} r_m^2.$$

Q is chi-square distributed with $k - p - q$ degrees of freedom.

The standard error of the partial autocorrelation is the same for all lags:

$$\text{S.E.}(\hat{\phi}_k) = \frac{1}{\sqrt{T}}.$$

4.1.2. The Inverse Autocorrelation. The inverse autocorrelation of an ARMA(p, q) process is defined as autocorrelation of the corresponding ARMA(q, p) process (see Cleveland, 1972). The inverse autocorrelation has a role similar to that of the partial autocorrelation and can be understood as the partial autocorrelation's alternative or complement. To identify seasonal, nonstationary, and almost nonstationary models, the inverse autocorrelation is often more helpful than the partial auto-correlation.

4.1.3. Extended Autocorrelation. The ACF and PACF most often clearly identify pure AR or pure MA models. ARMA models are less

m

```
    0 1 2 3 4 5 6
   ─────────────────
 0 │x x x x x x x
 1 │x x x x x x x
 2 │x 0 0 0 0 0 0
 3 │x x 0 0 0 0 0
k  4 │x x x 0 0 0 0
 5 │x x x x 0 0 0
 6 │x x x x x 0 0
 7 │x x x x x x 0
```

FIGURE 10.4.   ESACF indicator table for a special ARMA-model.

easily identified and differentiated. For that reason a number of attempts have been made to more clearly differentiate ARMA models. One example is the extended (sample) autocorrelation function (E(S)ACF), which proceeds in such a way that for $k = 1, 2, \ldots$ successive AR(k) models are determined and the $\phi_1, \ldots, \phi_k$ estimated. Residuals are formed each time (see Tsay and Tiao, 1984), and autocorrelations, denoted $r(k, m)$, are determined at lag m. These autocorrelations are the EACF.

$$r(k, m) \approx \begin{cases} c, & m = q + (k - p), \\ 0, & m > q + (k - p). \end{cases}$$

The EACF values can be summarized in condensed form by replacing those values that lie within two standard errors of zero with 0, x otherwise. The order of the ARMA model, p and q, can be determined by using this "indicator table" (see Figure 10.4). One searches in the lower right part of the matrix for a triangle containing only 0. The coordinates of the upper left tip of the triangle determine p (horizontal) and q (vertical). In the example here, the ESACF determines the triangle with tip indices $(2, 1)$. That means $p = 2$ and $q = 1$. Hence, an ARMA(2, 1) model is involved.

4.1.4.   Semiautomatic Identification with the Help of Criteria AIC, BIC, and HQ.   The identification of an "optimal" model is not always very simple, even with the procedures explained in the previous section. Subjective decisions must still be made at many points. It is often helpful to have a criterion by which various models can be compared with one another. A simple criterion available to the researcher is the residual variance s_a^2. The smaller the residual variance, and consequently the larger the variance explained by the model, the better the model fits.

The three most important model-fitting criteria are the *Akaike information criterion* (AIC)

$$\text{AIC}(p, q) = \ln \hat{\sigma}_{p,q}^2 + \frac{2(p + q)}{T}$$

The *Schwarz-Bayes criterion* (SBC), usually referred to as the *Bayes information criterion* (BIC) in the literature:

$$\text{SBC}(p, q) = \ln \hat{\sigma}_{p,q}^2 + \frac{(p + q)\ln(T)}{T} = \text{BIC}(p, q),$$

and the *Hannan-Quinn criterion* HQ

$$\text{HQ}(p, q) = \ln \hat{\sigma}_{p,q}^2 + 2(p + q)\frac{c \ln(\ln(T))}{T}, \qquad c > 1.$$

See Akaike (1976), Schwarz (1978), and Hannan and Quinn (1979) for derivation of the AIC, SBC, and HQ.

It has been shown that the BIC and the HQ are consistent, whereas the AIC is not consistent. In many cases AIC overestimates the true order of the model, but does not underestimate it if T is large enough. One should bear in mind that a schematic treatment of such criteria can cause one to overlook problems due to outliers or disregarded control variables, for example.

4.2. Estimation

When the form of a model has been identified, the parameters ϕ and θ must be estimated. Two methods often used in conventional statistics can also be used here: least squares and maximum likelihood. Instead of the exact maximum likelihood method, a *conditional* likelihood method is often used (see Box and Jenkins, 1976). The calculation of the *exact* likelihood function is very complex and tedious. The interested reader should refer to Ansley (1979) and Hannan, Dunsmuir, and Deistler (1980). According to these authors the calculation by means of exact maximum likelihood is rather time-consuming but is more accurate in many cases, especially when one is dealing with short series, models with many parameters, or models lying at the bounds of stationarity or invertibility.

4.3. Diagnostic Checking

With diagnostic checking, the specified model is examined for its appropriateness. Tools used to check the model are the significance testing of the residuals for white noise and of the ARIMA parameters. The residuals of an ARIMA model are a_t:

$$z_t - (\phi_1 z_{t-1} + \cdots + \phi_p z_{t-p} - \theta_1 a_{t-1} - \cdots - \theta_q a_{t-q}) = a_t$$

or

$$\text{Series} - \text{model} = \text{residuals}.$$

In keeping with the assumptions of the model, residuals should be white noise. To check the white noise assumption, the plot of the residual series must first be analyzed graphically. Second, the autocorrelation and partial autocorrelations of residuals must be tested for significance. In addition, the ARMA parameters should be tested for significance. These tests may indicate the need for more or less model parameters.

4.4. An Example of Empirical Time-Series Analysis [2]

The empirical data presented here are from a study of women coping with their first pregnancy. The theoretical background builds a process model of pregnancy and the transition to parenthood (see Gloger-Tippelt, 1983). The model emphasizes the phases when the mother begins to perceive the child as a separate being and when she accepts her future role as mother. It is surmised that the central variable—mother's evaluation of the pregnancy ("pregeva")—is highly dependent on her perceptions of physical changes, including the child's movements. Measurements were taken once a day over a period of 82 days. Our analysis concentrates on a single individual taken from this sample. The main concern here is to investigate the relationship between the evaluation of pregnancy and the evaluation of physical changes, done as multivariate time series analysis for these variables (see Section 5.4.2). Here we demonstrate the univariate analysis of the variable pregeva, not to answer substantive questions, but to facilitate the multivariate analyses, see Schmitz (1989). The identification of the univariate time-series model often can give important hints for modeling the multivariate time-series.

IDENTIFICATION. Figure 10.5 represents the series graphically. Figure 10.6 contains autocorrelation, inverse autocorrelation, and partial auto-

---

[2] We thank Mrs. Gloger-Tippelt, who made it possible to analyze this data set.

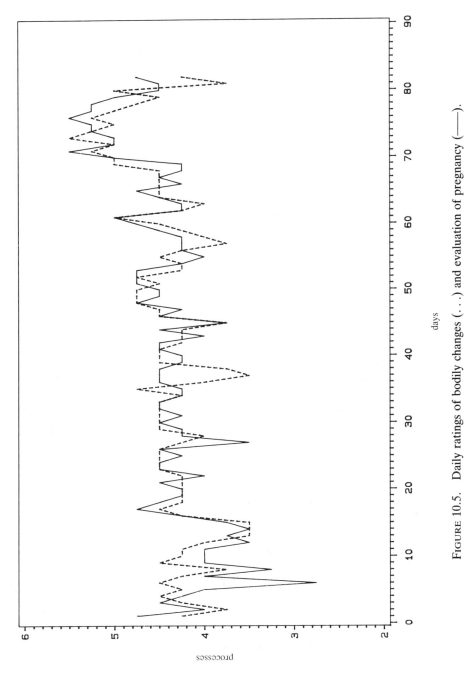

FIGURE 10.5.   Daily ratings of bodily changes (. . .) and evaluation of pregnancy (——).

```
AUTOCORRELATIONS

LAG COVARIANCE CORRELATION-1 9 8 7 6 5 4 3 2 1 0 1 2 3 4 5 6 7 8 9 1          STD
  0   0.220962    1.00000   |                    |********************|            0
  1   0.132547    0.59986   |              .     |************        |      0.110432
  2   0.126188    0.57108   |               .    |***********         |      0.144816
  3   0.0879472   0.39802   |                .   |*******            |       0.170077
  4   0.0920733   0.41669   |                .   |********           |        0.18108
  5   0.0649494   0.29394   |                .   |******   .         |       0.192419
  6   0.075991    0.34391   |                .   |*******.           |       0.197819
  7   0.056173    0.25422   |                .   |*****    .         |        0.20498
  8   0.0467655   0.21164   |                .   |****     .         |        0.20879
  9   0.0265014   0.11994   |                .   |**       .         |        0.21139
 10   0.0217971   0.09865   |                .   |**       .         |       0.212218
 11   0.0155685   0.07046   |               .    |*        .         |       0.212777
 12   0.0261082   0.11816   |               .    |**       .         |       0.213061
                            '.' MARKS TWO STANDARD ERRORS

INVERSE AUTOCORRELATIONS
         LAG CORRELATION -1 9 8 7 6 5 4 3 2 1 0 1 2 3 4 5 6 7 8 9 1
           1   -0.29835   |            ******|     .           |
           2   -0.15540   |             .***|     .           |
           3    0.03280   |               . |*    .           |
           4   -0.10923   |               . **|    .           |
           5    0.14439   |               . |***.             |
           6   -0.10176   |               . **|    .           |
           7   -0.03969   |               .  *|    .           |
           8   -0.01008   |               .   |    .           |
           9    0.01801   |               .   |    .           |
          10    0.07071   |               .   |*   .           |
          11    0.04959   |               .   |*   .           |
          12   -0.05647   |               .  *|    .           |

PARTIAL AUTOCORRELATIONS
         LAG CORRELATION -1 9 8 7 6 5 4 3 2 1 0 1 2 3 4 5 6 7 8 9 1
           1    0.59986   |               .   |************     |
           2    0.32999   |               .   |*******         |
           3   -0.05164   |               .  *|    .           |
           4    0.14169   |               .   |***.            |
           5   -0.04504   |               .  *|    .           |
           6    0.12095   |               .   |**  .           |
           7   -0.01603   |               .   |    .           |
           8   -0.08802   |               . **|    .           |
           9   -0.05995   |               .  *|    .           |
          10   -0.02965   |               .  *|    .           |
          11    0.02910   |               .   |*   .           |
          12    0.09912   |               .   |**  .           |

                    AUTOCORRELATION CHECK FOR WHITE NOISE
         TO    CHI                      AUTOCORRELATIONS
     LAG  SQUARE DF  PROB
       6  106.27  6  0.000   0.600  0.571  0.398  0.417  0.294  0.344
      12  120.52 12  0.000   0.254  0.212  0.120  0.099  0.070  0.118
      18  123.97 18  0.000   0.105  0.095  0.058  0.066  0.048  0.060
      24  126.95 24  0.000   0.075  0.046  0.085  0.029  0.103  0.009
```

FIGURE 10.6. Identification results: autocorrelations, partial autocorrelations and inverse autocorrelations Ljung-Box test for the series "pregeva."

TABLE 10.1.   Results of the Estimation and Diagnostic Checking
for the Series

Parameter	Estimate	Approx. std. error	T-ratio	Lag
MU	4.41326	0.138074	31.96	0
AR1, 1	0.398141	0.105626	3.77	1
AR1, 2	0.326995	0.106067	3.08	2
AIC	= 69.0672			
SBC	= 76.2873			

Autocorrelation check of residuals

To lag	Chi-square	DF	Prob	Autocorrelations				
6	3.87	3	0.276	0.027	−0.014	−0.101	0.079	−0.066	0.147
12	6.44	9	0.696	0.079	0.048	−0.052	−0.070	−0.070	0.087
18	7.62	15	0.938	0.086	0.033	−0.044	0.001	−0.032	0.017
24	11.08	21	0.961	0.048	−0.027	0.042	−0.057	0.136	−0.056

correlation functions, plots with significance limits, and the Ljung-Box
Q-test. Note that the autocorrelations decrease exponentially and the
partial autocorrelations vanish after two lags, clearly indicating that an
AR(2) model is involved. The behavior of the inverse autocorrelations
(having similar cutoff properties as the partial autocorrelations) shows a
decay after one or two lags and are also compatible with an AR(2)
process.

```
                          AUTOCORRELATION PLOT OF RESIDUALS

LAG COVARIANCE CORRELATION-1 9 8 7 6 5 4 3 2 1 0 1 2 3 4 5 6 7 8 9 1        STD
  0    0.130093    1.00000   |                    |********************|        0
  1  0.00354275    0.02723   |                    .  |*  .              |  0.110432
  2 -0.0018671    -0.01435   |                    .  |   .              |  0.110513
  3 -0.0131167    -0.10083   |                    . **|   .              |  0.110536
  4   0.010231     0.07864   |                    .  |** .              |  0.111652
  5 -.00857006    -0.06588   |                    .  *|   .              |  0.112326
  6  0.0191635     0.14731   |                    .  |***.              |  0.112796
  7  0.0102423     0.07873   |                    .  |** .              |  0.115118
  8 0.00629015     0.04835   |                    .  |*  .              |  0.115773
  9 -.00737485    -0.05669   |                    .  *|   .              |  0.116019
 10 -0.0068094    -0.05234   |                    .  *|   .              |  0.116356
 11 -.00908163    -0.06981   |                    .  *|   .              |  0.116643
 12  0.0113047     0.08690   |                    .  |** .              |  0.117151
```

FIGURE 10.7.   Diagnostic checking for the series "pregeva."

ESTIMATION.   Based on the examination of these functions, we estimate and AR(2) model. Table 10.1 and Figure 10.7 show the results. Both AR parameters are significant. Therefore, an AR(2) model is indicated. All statistics are consistent with the interpretation that the residual time series is white noise. An AR model of order 2 means that the variable pregnancy evaluation at time t can best be predicted from two previous time points. There is no need to include seasonal parameters.

5.   Multivariate ARIMA Model

Multivariate models permit the dynamic analysis of variables' inter-relationships. Because variables are temporally ordered, hypotheses about the direction of influences can be tested.

5.1.   Basic Terms

5.1.1.   Cross-Correlation.   Correlations between two time series are called *cross-correlations*. The synchronic correlations between two time series give no clue as to the direction of possible mutual influence. By contrast, an asynchronic cross-correlation is an indicator of unilateral influence. Under certain circumstances, lagged influences between two time series can be present in both directions. In such cases one speaks of *feedback*. Unfortunately, if individual time series show serial dependencies, cross-correlations do not always provide an accurate picture of the underlying causal relationships. Possible corrections for these correlations are discussed later. Multivariate ARIMA models (see Section 5.2) provide better insight into dynamic relations between variables than do cross-correlations.

5.1.2.   Wiener–Granger Causality.   The concept of causality used in this chapter stems from the work of Wiener (1956) and Granger (1969), and is often discussed in the literature as Wiener–Granger causality or Granger causality. The following treatment is based largely on Kirchgässner (1981). The variance of the forecasting error relating to the prediction of y_{t+k} under the condition I_t is understood to be

$$\sigma^2(y_{t+k} \mid I_t).$$

The definition of Wiener–Granger causality (as in Kirchgässner, 1981, pp. 21–22) is that x is *causal* in relation to y if and only if

$$\sigma^2(y_{t+1} \mid I_t) < \sigma^2(y_{t+1} \mid I_t - x_{\bar{t}}),$$

where

$$x_{\bar{t}} := (x_t, x_{t-1}, \dots).$$

This means that y_{t+1} can be predicted more accurately with $(x_t, x_{t-1}, \dots)$ than without $(x_t, x_{t-1}, \dots)$. Since this definition uses the term *causality* (although it has to do only with predictability), we will use the term *lagged incremental predictability* (LIP). "Lagged" means that y_{t+1} is predicted by x_t, x_{t-1}, etc. *Simultaneous incremental predictability* (SIP) exists for y by x if

$$\sigma^2(y_{t+1} \mid I_t, x_{t+1}) < \sigma^2(y_{t+1} \mid I_t).$$

This means that y_{t+1} can be predicted better with x_{t+1} than without x_{t+1}. Instead of the term simultaneous incremental predictability, sometimes the term *instantaneous incremental predictability* is used. With the help of LIP, feedback can now be defined. *Feedback* is present if there exists LIP (x by y) and LIP (y by x).

Note that LIP and, consequently, feedback are defined as functions of the choice of time lag. LIP and feedback might exist in terms of one time grid but not of another.

5.1.3. Possible Relations between Two Variables.

In addition to the relationships already pointed out, the mutually exclusive relations pre-

TABLE 10.2. Disjunct Predictability Relationships for Two Time Series x and y

There exists no SIP.	
x, y are independent	(x, y)
x is only causal to y	$(x \rightarrow y)$
y is only causal to x	$(x \leftarrow y)$
There is feedback	$(x \leftrightarrow y)$
There exists SIP.	
There is only SIP	$(x - y)$
x is only causal to y and SIP	$(x \Rightarrow y)$
y is only causal to x and SIP	$(x \Rightarrow y)$
There is feedback and SIP	$(x \Leftrightarrow y)$

sented in Table 10.2 are also possible between two time-series variables x and y (see Kirchgässner, 1981).

5.1.4. Using Cross-Correlations to Determine Predictability Relations.
Like autocorrelations, cross-correlations are functions of the lag that exists between series. Unlike autocorrelations, however, the direction of the shift is important.

The cross-correlation to lag k is defined as

$$\sigma_{ij}(k) = E((x_{it} - \mu_i)(x_{jt-k} - \mu_j)) = E(z_{it} \cdot z_{jt-k}),$$

$$\rho_{ij}(k) = \frac{\sigma_{ij}(k)}{(\mathrm{var}(z_{it}) \, \mathrm{var}(z_{jt}))^{1/2}}.$$

A consistent estimate is provided by

$$r_{ij}(k) = \frac{\sum_{t=k+1}^{T} (x_{it} - \bar{x}_i)(x_{jt-k} - \bar{x}_j)}{(\sum_{t=1}^{T} (x_{it} - \bar{x}_i)^2 \sum_{t=1}^{T} (x_{jt} - \bar{x}_j)^2)^{1/2}}.$$

The second index j indicates the earlier series, that is, the one relating to lag k if k is positive. The autocorrelation is symmetrical:

$$\rho_k = \rho_{-k}.$$

The formula for the cross-correlation is

$$\rho_{ij}(k) = \rho_{ji}(-k).$$

When using cross-correlations to help analyze relations between time series, one must *not* test the cross-correlations for significance in the manner customary for conventional correlations. Cross-correlations are influenced by the serial dependence of the individual time series. Thus, relationships between the time series can appear stronger than they really are, showing a *spurious correlation*. The nature and extent of this distortion can be found in a simulation study by Schmitz (1987). Interpreting cross-correlations can therefore be misleading if the internal structures of the univariate series are not taken into consideration.

There are two appropriate ways to take serial dependence into consideration:

1. Using the Bartlett test (see Box and Jenkins, 1976, p. 377).
2. Filtering the serial dependence out of the univariate time series (prewhitening) and calculating cross-correlations between filtered series. Specific patterns of the residuals' cross-correlations characterize specific predictability relations.

TABLE 10.3.   The Identification of Predictability Relationships
Using Cross-Correlations of Residuals a_t, b_t

If there exists no SIP:

$$(x \to y) \Leftrightarrow (\exists k, k > 0: \rho_{ab}(k) \neq 0) \text{ and } (\forall k, k \leq 0: \rho_{ab}(k) = 0)$$

$$(x \leftarrow y) \Leftrightarrow (\exists k, k < 0: \rho_{ab}(k) \neq 0) \text{ and } (\forall k, k \geq 0: \rho_{ab}(k) = 0)$$

$$(x \leftrightarrow y) \Leftrightarrow (\exists k_1, k_1 < 0: \rho_{ab}(k_1) \neq 0) \text{ and } (\exists k_2, k_2 > 0: \rho_{ab}(k) \neq 0)$$

Table 10.3, adapted from Kirchgässner (1981, p. 33), gives correspondence rules for some of the results of Table 10.2.

In order to identify $x \to y$, at least one positive lag must be found for which the residual correlation is different from 0, while the cross-correlations of the residuals must be 0 for all negative and 0 lags. Individual cross-correlations as well as groups of cross-correlations can be tested for significance (see Mark, 1979; Pierce and Haugh, 1977). Pre-whitening is not free of problems, either (see also Schmitz, 1987). First, note that an assumption is made that *no* simultaneous predictability exists. In addition, the identification of an ARIMA model is not always unequivocal. Hence, it is conceivable that an inappropriate univariate ARIMA model can lead to erroneous estimates. Also, in many studies for which theoretical considerations lead to strong and well-founded assumptions about LIP between two variables, empirically investigated lagged cross-correlations have been shown to not exceed the level of statistical significance. Schwert (1979) has proven that the power of the Haugh-Pierce test, which builds on residual cross-correlations, is relatively low. This means, in some cases, true existing relationships will not be detected. This is the phenomenon of *spurious independence,* where relations that actually do exist are obscured. Finally, cross-correlations investigate the relationship between only *two* variables.

5.2.   Multivariate ARIMA Models

The multivariate ARIMA models by Jenkins and Alavi (1981) and Tiao and Box (1981) are, in many ways, preferable to the cross-correlations. Multivariate ARIMA models avoid problems of spurious correlation and of spurious independence. The first applications of such models in psychology were reported by Keeser and Bullinger (1984) and Gregson (1984). To avoid detailing unnecessarily lengthy and complicated formulas here, the following presentation is restricted to the description of bivariate processes.

5.2.1. White Noise. Bivariate white noise processes consist of two univariate white noise processes a_{1t} and a_{2t}, each with an expectation of 0:

$$E(\mathbf{a}_t) = \begin{pmatrix} 0 \\ 0 \end{pmatrix}.$$

Both processes can covary simultaneously:

$$E(a_{1t}a_{2t}) = \text{cov}(a_{1t}, a_{2t}) = \sigma_{12}.$$

But for all lags not equal to 0, there is no relationship between the two processes. Usually it is assumed that the two variables are bivariately normally distributed.

$$\mathbf{a}_t \sim N(\mathbf{0}, \Sigma), \quad E(\mathbf{a}_t a'_{t-k}) = \mathbf{0}, \qquad k \neq 0.$$

5.2.2. AR Models. From a univariate AR(p):

$$z_t = \phi_1 z_{t-1} + a_t,$$

one formally derives the multivariate AR(p) process by replacing relevant scalars with vectors or matrices. The formula for the multivariate AR(1) process is

$$\begin{pmatrix} z_{1t} \\ z_{2t} \end{pmatrix} = \begin{pmatrix} \phi_{11} & \phi_{12} \\ \phi_{21} & \phi_{22} \end{pmatrix} \begin{pmatrix} z_{1t-1} \\ z_{2t-1} \end{pmatrix} + \begin{pmatrix} a_{1t} \\ a_{2t} \end{pmatrix}.$$

For the multivariate AR(2) process the formula is

$$\begin{pmatrix} z_{1t} \\ z_{2t} \end{pmatrix} = \begin{pmatrix} \phi_{11}^1 & \phi_{12}^1 \\ \phi_{21}^2 & \phi_{22}^2 \end{pmatrix} \begin{pmatrix} z_{1t-1} \\ z_{2t-2} \end{pmatrix} + \begin{pmatrix} \phi_{11}^2 & \phi_{12}^2 \\ \phi_{21}^2 & \phi_{22}^2 \end{pmatrix} \begin{pmatrix} z_{1t-2} \\ z_{1t-2} \end{pmatrix} + \begin{pmatrix} a_{1t} \\ a_{2t} \end{pmatrix}.$$

In order to better understand the multivariate AR(1) model, we distinguish important special cases of this model:

1. Phi matrix is **0**.
2. Phi matrix is diagonal.
3. Phi matrix is triangular.
    3.1. Phi matrix is upper triangular.
    3.2. Phi matrix is lower triangular.
4. Phi matrix has a complete off-diagonal.

Phi's diagonal elements give information about the autodependencies of the univariate processes, and phi's off-diagonal elements give information about the lagged relations between processes. Information about simultaneous relations between variables is given by the variance-covariance matrix of the residuals, sigma. If nonzero elements exist in the off-diagonals of sigma, then simultaneous relationships between variables exist. Given these generalizations, consider the special cases mentioned earlier:

1. Phi matrix is **0**. If all $\phi_{i,j} = 0$, then the AR(1) model corresponds to white noise. Absolutely no lagged patterns are present.

2. Phi matrix is diagonal. If ϕ is a diagonal matrix, two univariate AR(1) processes are involved, with no lagged relations between them.

3. Phi matrix is triangular. If phi is an upper- (or a lower-)triangular matrix, that is, if $\phi_{1,2} \neq 0$ (or $\phi_{2,1} \neq 0$), and if parameters different from 0 are only above (or below) and possibly also on the diagonal, then a special transfer function model is involved:

$$\begin{pmatrix} z_{1t} \\ z_{2t} \end{pmatrix} = \begin{pmatrix} \phi_{11} & \phi_{12} \\ 0 & \phi_{22} \end{pmatrix} \begin{pmatrix} z_{1t-1} \\ z_{2t-1} \end{pmatrix} + \begin{pmatrix} a_{1t} \\ a_{2t} \end{pmatrix}.$$

z_{1t} is influenced by z_{1t-1}, z_{2t-1}. By contrast, z_{2t} is influenced at most by z_{2t-1}. Hence, there is a unidirectional relationship. Quite generally, one speaks of a *transfer-function* model if the influence of an input on an output is being studied or, in several processes, if every variable is either input or output, but not both at the same time. In the multivariate model, it need not be specified a priori which variable is to be considered an input and which an output. The ARIMA model coefficients indicate which of the different patterns is involved.

TABLE 10.4. The Identification of Predictability Relationships for a Multivariate AR(1)-Model

$(z_1, z_2) \Leftrightarrow \phi_{12}, \phi_{21} = 0$
$(z_1 \rightarrow z_2) \Leftrightarrow \phi_{21} \neq 0, \phi_{12} = 0$
$(z_2 \rightarrow z_1) \Leftrightarrow \phi_{12} \neq 0, \phi_{21} = 0$
$(z_1 \leftrightarrow z_2) \Leftrightarrow \phi_{12}, \phi_{21} \neq 0$

4. Phi matrix has a complete off-diagonal. If neither $\phi_{1,2}$ nor $\phi_{2,1}$ equals 0, then lagged relations exist between z_1 and z_2 as well as between z_2 and z_1; *feedback* is present.

Table 10.4 summarizes the most important patterns in their connection to the predictability relations for a multivariate AR(1) model.

The relationship between the multivariate and the univariate models of the component time series depends on the kind of predictability relation of the multivariate AR(1) model. If there exists feedback, for example, the maximum order of the ARMA model for both variables can be $p = 2$ and $q = 1$. If there exists no LIP, the maximum order for both univariate series is $p = 1$, $q = 0$. For further details see Jenkins and Alavi (1981).

5.2.3. MA Models. For MA models, too, the generalization from univariate to multivariate case is very simple. A multivariate MA(1) may be written

$$\begin{pmatrix} z_{1t} \\ z_{2t} \end{pmatrix} = \begin{pmatrix} a_{1t} \\ a_{2t} \end{pmatrix} - \begin{pmatrix} \theta_{11} & \theta_{12} \\ \theta_{21} & \theta_{22} \end{pmatrix} \begin{pmatrix} a_{1t-1} \\ a_{2t-1} \end{pmatrix}.$$

The general multivariate MA(q) process is defined accordingly. The current component processes z_{1t} and z_{2t} are derived from current white noise components, a_{1t} and a_{2t}, and the weighted white noise parts of their own process and of the other process, a_{1t-1} and a_{2t-1} at a previous moment. The classification of the predictability relations vis-à-vis the parameters θ and Σ corresponds to that of the AR(1) process when corresponding statements about ϕ are replaced by θ.

5.2.4. Autocorrelation and Partial Autocorrelation. The auto-covariance Γ and autocorrelation ρ at lag k for a times series z_t are defined as

$$\Gamma(k) = \begin{pmatrix} \sigma_{11}(k) & \sigma_{12}(k) \\ \sigma_{21}(k) & \sigma_{22}(k) \end{pmatrix},$$

$$\rho(k) = \begin{pmatrix} \rho_{11}(k) & \rho_{12}(k) \\ \rho_{21}(k) & \rho_{22}(k) \end{pmatrix}.$$

For σ_{ij}, ρ_{ij}, see 5.1.4. The autocovariance or autocorrelation of a multivariate time series is found similarly.

The multivariate partial autocorrelation $\mathbf{P}(k)$ is defined as the last coefficient in the multivariate autoregressive regression equation containing k predictors.

$$\mathbf{z}_t = \boldsymbol{\phi}_1 \mathbf{z}_{t-1} + \cdots + \boldsymbol{\phi}_k \mathbf{z}_{t-k} + \mathbf{a}_t,$$

$$\mathbf{P}(k) = \boldsymbol{\phi}_k (= \boldsymbol{\phi}_{kk}) = \begin{pmatrix} p_{11}(k) & p_{12}(k) \\ p_{21}(k) & p_{22}(k) \end{pmatrix}.$$

The patterns of the multivariate ACF and PACF match those of the univariates completely. With a MA(q), the autocorrelation cuts off after q lags; with an AR(p), the partial autocorrelation cuts off after p lags.

5.2.5. ARMA Models.   The general ARMA model has the form

$$\boldsymbol{\phi}_p(B)\mathbf{z}_t = \boldsymbol{\theta}_q(B)\mathbf{a}_t,$$

$$\boldsymbol{\phi}_p(B) = (I - \boldsymbol{\phi}_1 B^1 - \cdots - \boldsymbol{\phi}_p B^p),$$

$$\boldsymbol{\theta}_q(B) = (I - \boldsymbol{\theta}_1 B^1 - \cdots - \boldsymbol{\theta}_q B^q).$$

Parallel conditions of stationarity and invertibility for multivariate ARMA models exist (see Tiao and Box, 1981).

5.3.   Analyzing Multivariate Time Series

Just as with univariate ARIMA models, multivariate ARIMA models are determined through the iterative procedure of identification, estimation, and diagnosis.

5.3.1. Identification.   As with univariate models, multivariate models are identified by comparing the patterns of empirical time series' autocorrelation and partial autocorrelation with those of known models.
     The matrix $\mathbf{R}(k)$ gives the sample autocorrelation at lag k:

$$\mathbf{R}(k) = \begin{pmatrix} r_{11}(k) & r_{12}(k) \\ r_{21}(k) & r_{22}(k) \end{pmatrix}.$$

The check

$$|r_{ij}(k)| < 2\left(\frac{1}{\sqrt{T}}\right),$$

which assumes a white noise process, should by no means be thought of as a formal test of significance. It can serve only as a rough rule of thumb. One may also heuristically examine autocorrelations with the help of *indicator symbols* $(+, -,$ or $\cdot)$ based on this rule of thumb, which gives an effective summary of the autocorrelation pattern.

$$r_{ij}(k) > \frac{2}{\sqrt{T}}, \qquad +,$$

$$r_{ij}(k) < \frac{-2}{\sqrt{T}}, \qquad -,$$

$$\frac{-2}{\sqrt{T}} \leq r_{ij}(k) \leq \frac{2}{\sqrt{T}}, \qquad \cdot\;.$$

Sample partial autocorrelations result when the regression equation (2) is estimated for the empirical time series.

A further aid for tentative identification of an autoregressive model is the $M(m)$ statistic (see Tiao and Box, 1981). With this statistic, the null hypothesis tested is $\phi_m = 0$ as opposed to the alternative $\phi_m \neq 0$, with ϕ_m being the mth coefficient in an autoregressive $AR(m)$ model. Assuming the validity of the null hypothesis, $M(m)$ is asymptotically distributed according to a chi-square with d^2 degrees of freedom, with d the dimension of the time series.

The model selection criteria presented for the univariate case can be generalized as well (see Lütkepohl, 1985; Priestley, 1981, p. 690).

5.3.2. Estimation. After the order of the model (p, q) has been tentatively specified, parameters can be estimated. There are procedures that build upon the conditional (Wilson, 1973) as well as upon the exact maximum likelihood function (Hillmer and Tiao, 1979).

5.3.3. Diagnostics. To detect deviations from the model, it is advisable to plot every univariate residual series in standardized form. Since a properly fitting model will produce a residual series $\mathbf{a}_t$ that is multivariate white noise, calculated autocorrelations of the residuals may be tested to see if values significantly differ from 0. To take account of multiple statistical tests, the portmanteau test, analogous to the Ljung-Box Q-test, may be used. The portmanteau test checks a series of correlation matrices for white noise simultaneously (see Hosking, 1980). An analysis of the residual series can also include the testing of partial correlations, $M(m)$ statistics, and the diagonals of the variance-covariance matrix of the residuals.

5.4. Examples of How to Conduct Multivariate Time-Series Analyses

5.4.1. Multivariate Analyses Using Simulated Data

IDENTIFICATION.   Figure 10.8 shows the trajectory of the two-dimensional time-series vector (V1, V2). This simulation includes 100 data points, cf. Schmitz (1989). Table 10.5 gives the results.

The correlation between the two series (−.29) is significant. The correlation matrix at lag 1 is the only matrix that contains nothing but nonzero values. The autocorrelation function thus cuts off after lag 1, indicating an MA(1) process. Among other things, Table 10.6 lists the residual variances for each lag, the $M(m)$ statistic (denoted here as a chi-square test), the AIC, and the significance of the partial correlations. The $M(m)$ statistic is significant up to lag 2, the minimal AIC is −.068 at lag 2, and even the partial correlation does not vanish until after lag 2. Taken together, the pattern of cross-correlations and partial correlations indicate a MA(1) model.

ESTIMATION.   In the second step we assert that an MA(1) model is involved. Table 10.6 shows the results of the estimation of this model. In the θ matrix, all parameters are significant.

DIAGNOSTIC CHECKING.   In addition to estimation parameters, Table 10.6 contains diagnostic output. There are conspicuous cross-correlations for lag 5 only. Similarly, except for isolated values for lag 5, the partial correlations are also insignificant. On the whole, the obtained residual series is consistent with the characteristics of white noise. The assumption that an MA(1) model is involved can therefore be accepted.

The estimated model describes a bidirectional predictability relation (feedback) in which V1 has lagged influence on V2, and vice versa, and in which there is no simultaneous predictability. The correlation of the residuals is not significant.

5.4.2. Multivariate Analyses Using Empirical Data: The Relationship Between Evaluation of Pregnancy and Evaluation of Physical Changes.   For a short description of the aims of the study and of previous analyses, see Section 4.4. Figure 10.5 shows the time course of the variables. From the fact that the univariate analysis of the variable pregeva follows an AR(2), it can be concluded that in case of an

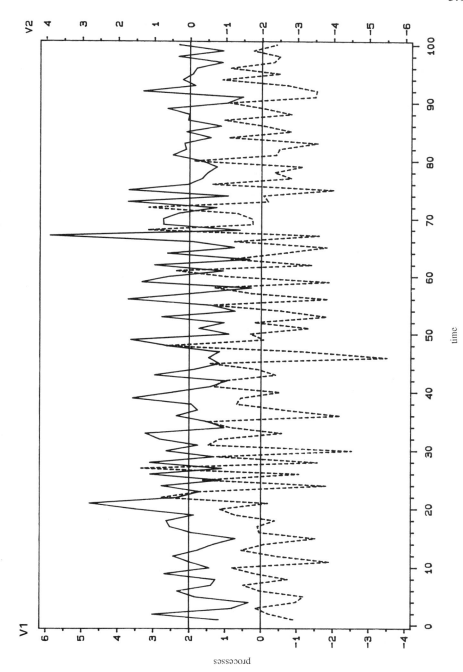

FIGURE 10.8. Two simulated time series: V1 (. . ., scaling at the left axis) and V2 (——, scaling at the right axis).

TABLE 10.5. Multivariate Identification Results: Cross-Correlations and Partial Correlations for the Simulated Series

Lag	Residual variances	Chi-square test	AIC	PACF indicators	ACF indicators
1	1.38	33.86	.127	· +	− +
	.79			+ ·	+ −
2	1.12	24.38	−.068	− +	· ·
	.72			· −	· ·
3	1.07	3.62	−.030	· ·	· ·
	.72			· ·	· ·
4	1.04	3.69	.006	· ·	· ·
	.72			· ·	· ·
5	.97	5.18	.023	· ·	· ·
	.72			· ·	· ·
6	.90	8.48	−.002	· +	· ·
	.68			· ·	· ·

Note: Chi-squared critical values with four degrees of freedom are 5%: 9.5; 1%: 13.3.

TABLE 10.6. Results of the Multivariate Estimation and Residual Analysis for the Simulated Series

Estimates of theta(1) *matrix, significance, and standard errors*

.452	−.827	+ −	.077	.089
−.338	.265	− +	.062	.071

Lag	Residual variances	Chi-square test	AIC	PACF indicators	ACF indicators
1	.95	4.78	−.263	· ·	· ·
	.75			· ·	· ·
2	.92	6.30	−.257	· ·	· ·
	.72			· ·	· ·
3	.92	.97	−.189	· ·	· ·
	.71			· ·	· ·
4	.90	2.13	−.136	· ·	· ·
	.70			· ·	· ·
5	.78	12.43	−.214	− +	− +
	.69			· ·	· ·
6	.78	2.20	−.163	· ·	· ·
	.68			· ·	· ·

Note: Chi-squared critical values with four degrees of freedom are 5%: 9.5; 1%: 13.3.

multivariate AR(1) model would be appropriate, there must be at least one LIP relationship. Table 10.7 shows the results. The cross-correlation of .63 at lag 0 is high. The pattern of cross-correlations remains stable up to lag 7. The PACF cuts off after lag 1, although a single value is also conspicuous at lag 2. $M(m)$ chi-square indicates an AR(1); AIC, an AR(2) model. For the time being, we tentatively estimate an AR(1) model (see Table 10.8).

The estimate provides a ϕ matrix in which the parameters in the upper triangle are significant. The residual cross-correlations conform to white noise. The parameter matrix shows a unidirectional relation. The evaluation of physical change (second component series) influences the evaluation of pregnancy (first component series). But the parameter for the reverse relation is on the very margin of significance $\phi_{2,1} = .20$ (S.E. = .11). It is therefore appropriate to summarize the results as follows: In addition to the simultaneous relation between variables $r = .42$ (S.E. = .11), the foregoing described unidirectional lagged influence (as well as possibly feedback) exists.

TABLE 10.7. Multivariate Identification Results: Cross-Correlations and Partial Correlations for the "Pregnancy" Example

Lag	Residual variances	Chi-square test	AIC	PACF indicators	ACF indicators Lags 1–6	ACF indicators Lags 7–12
1	.47	55.74	−1.426	+ +	+ +	+ +
	.60			+ +	+ +	· ·
2	.42	8.19	−1.445	+ ·	+ +	· ·
	.58			· ·	+ +	· ·
3	.42	1.97	−1.376	· ·	+ +	· ·
	.56			· ·	+ +	· ·
4	.40	2.99	−1.323	· ·	+ +	· ·
	.56			· ·	+ +	· ·
5	.40	1.52	−1.249	· ·	+ +	· ·
	.56			· ·	· +	· ·
6	.39	1.95	−1.183	· ·	+ +	· ·
	.55			· ·	+ +	· ·

Note: Chi-squared critical values with four degrees of freedom are 5%: 9.5; 1%: 13.3.

TABLE 10.8. Results of the Multivariate Estimation and Residual Analysis for the "Pregnancy" Example

Estimates of phi(1) *matrix, significance, and standard errors*				
.434	.267	+ +	.111	.110
.202	.483	· +	.112	.111

Residual analysis

Sample correlation matrix of the series

1.00		The approximate std. error is
0.42	1.00	$(1/T^{**}.5) = 0.11$

Lags 1 through 6

··	+ ·	··	··	··	··
··	· ·	··	··	··	··

The sign of the synchronous relationship shows that days with positive (negative) evaluation of pregnancy coincide with days of positive (negative) evaluation of bodily changes. But moreover, there is a time-lagged relationship, which indicates that days when the mother evaluates her bodily changes as positive are followed by days when she has a positive feeling about her pregnancy as well. Thus the thesis regarding the importance of the physical changes is confirmed. But with high probability there also is an inverse time-lagged relationship: The interpretation of physical changes is influenced by the general evaluation of the pregnancy. In sum, with high probability it can be concluded that there is an *intraindividual dynamic interaction* of processes.

6. Summary and Discussion

To analyze intraindividual variability and relationships, new developments in time-series methods were discussed. Of particular importance to psychology is the analysis of time-lagged relationships, because results allow conclusions regarding "causal" mechanisms. We prefer the term *lagged incremental predictability* (LIP) instead of *causality*. It was argued that for the analysis of LIP, multivariate ARIMA models are appropriate and are preferable to the analysis of cross-correlations. Although the ARIMA models provide powerful analytic tools, they are based on some crucial assumptions:

1. The correct time lag is specified.
2. The (differenced) time series must be stationary.
3. The models are linear; that is, the series can be represented as a linear combination of past values of the series and the random shocks.

The importance of the time lag has already been mentioned in the definition of LIP. Because LIP could be detected for one time lag but not for a greater one, in longitudinal research the appropriate time lag should be determined as a first step. This could be done in pilot (single-case) studies based on the finest possible lag. Starting with this finest lag, analyses with varying lags may be conducted to find for which lag the probability of LIP discovery is maximal.

As Priestley (1981) points out, the real world is neither linear nor stationary. It might be useful to have some methods that can detect at least special cases of nonstationarity. In the ARIMA model, stationarity is accomplished by suitable differencing. But this may be rather inappropriate because this procedure handles qualitatively different processes in the same way. For example, differencing does not differentiate between a time series that shows some level shifts and one that is trending. Schmitz (1987) has proposed a method that allows one to partition a nonstationary time series into homogeneous phases. Harvey's (1981) state-space models provide another way to deal with non-stationarity, which may be superior to differencing.

To handle nonlinearity, Tong (1981) has introduced the so-called threshold models. We give an example how these models could be applied: If the regulation (represented by AR-parameters) of mood will be different for high and low mood values, the mean mood level serves as threshold and the AR-parameters are modeled in dependency of the threshold. Another, more general, class of nonlinear models, state-dependent models (SDM), is proposed by Priestley (1987). Here the time series parameters can depend on the state of the process. For a more detailed description of these models see Schmitz (1989).

All the just mentioned nonstationary or nonlinear models can be viewed as generalizations of the ARIMA model. In this way studying the ARIMA models is a beginning to understanding analyses of time series.

In sum, it would be desirable if the predictions of Mark (1979) would be fulfilled: "Work on the causal analysis of time-series data will expand considerably in the coming years and that the methodological advances will be plentiful" (p. 321).

References

Akaike, H. (1976). "Canonical correlation analysis of time series and the use of an information criterion." In R. K. Lainiotis and D. G. Lainiotis (Eds.), *System identification.* New York: Academic Press.

Akaike, H. (1979). "A Bayesian extension of the minimum AIC procedure of autoregressive model fitting," *Biometrika.* **66,** 242–273.

✓ Allport, G. W. (1962). "The general and the unique in psychological science," *Journal of Personality.* **30,** 405–422.

Ansley, C. F. (1979). "An algorithm for the exact likelihood of a mixed autoregressive-moving average process," *Biometrika.* **66,** 59–65.

Baltes, P. B., and Willis, S. L. (1982). "Plasticity and enhancement of intellectual functioning in old age: Penn State's adult development and enrichment project (ADEPT)." In F. I. M. Craik and S. E. Trehub (Eds.), *Aging and cognitive processes* (pp. 353–389). New York: Plenum.

Baltes, P. B., Reese, H. W., and Nesselroade, J. R. (1977). *Life-span developmental psychology: introduction to research methods.* Monterey, Calif.: Brooks/Cole.

Box, G. E. P., and Jenkins, G. M. (1976). *Forecasting and control.* San Francisco: Holden-Day.

Cleveland, W. W. (1972). "The inverse autocorrelations of a time series and their applications," *Technometrics.* **14,** 277–293.

Glass, G. V., Willson, V. L., and Gottman, J. M. (1975). *Design and analysis of time-series experiments.* Boulder, Colo.: Associated University Press.

Gloger-Tippelt, G. (1983). "A process model of the pregnancy course," *Human Development.* **26,** 134–148.

Gottman, J. M. (1981). *Time-series analysis.* Cambridge, England: Cambridge University Press.

Granger, C. W. J. (1969). "Investigating causal relations by econometric models and cross-spectral methods," *Econometrica.* **37,** 424–438.

Gregson, R. A. M. (1983). *Time series in psychology.* Hillsdale, N.J.: Lawrence Erlbaum.

Gregson, R. A. M. (1984). "Invariance in time-series representations of 2-input 2-output psychophysical experiments," *British Journal of Mathematical and Statistical Psychology.* **37,** 100–121.

Hannan, E. J., and Quinn, B. G. (1979). "The determination of the order of an autoregression," *Journal of the Royal Statistical Society.* **41,** 190–195.

Hannan, E. J., Dunsmuir, W. T. M., and Deistler, M. (1980). "Estimation of vector armax models," *Journal of Multivariate Analysis.* **10,** 275–295.

Harris, Ch. W. (Ed.) (1963). *Problems in measuring change*. Madison, WI: University of Wisconsin Press.

Harvey, A. C. (1981). *Time-series models*. Oxford: Philip Alan.

Hillmer, S. C., and Tiao, G. C. (1979). "Likelihood function of stationary multiple autoregressive moving average models," *Journal of the American Statistical Association*. **74,** 652–661.

Hosking, J. R. M. (1980). "The multivariate portmanteau statistic," *Journal of the American Statistical Association*. **75,** 602–607.

Jenkins, G. M., and Alavi, A. S. (1981). "Modelling and forecasting multivariate time series," *Journal of Time Series Analysis*. **2,** 1–47.

Keeser, W., and Bullinger, M. (1984). "Process-oriented evaluation of a cognitive-behavioural treatment for clinical pain: a time-series approach." In B. Bromm (Ed.), *Pain measurement in man*. Amsterdam: Elsevier.

Kirchgässner, G. (1981). *Einige neuere statistische Verfahren zur Erfassung kausaler Beziehungen zwischen Zeitreihen*. Göttingen, W. Germany: Vandenhoeck and Ruprecht.

Kohlberg, L. (1963). "The development of children's orientation toward a moral order: I. Sequence in the development of moral thought," *Vita Humana*. **6,** 11–33.

Ljung, G. M., and Box, G. E. P. (1978). "On a measure of lack of fit in time-series models," *Biometrika*. **65,** 297–303.

Lütkepohl, H. (1985). "Comparison of criteria for estimating the order of a vector autoregressive process," *Journal of Time Series Analysis*. **6,** 35–52.

Mark, M. M. (1979). "The causal analysis of concomitancies in time series." In T. D. Cook and D. T. Campbell (Eds.), *Quasi-Experimentation*. Chicago: Rand McNally.

McCleary, R., and Hay, A. R. (1980). *Applied time series analysing*. Beverly Hills, Calif.: Sage.

Piaget, J. (1948). *Psychologie der Intelligenz*. Zürich: Rascher.

Pierce, D., and Haugh, L. D. (1977). "Causality in temporal systems, characterizations and a survey," *Journal of Econometrics*. **5,** 265–293.

Priestley, M. B. (1981). *Spectral analysis and time series*, Vols. I, II. New York: Academic Press.

Priestley, M. B. (1987). "New developments in time-series analysis." In M. L. Puri, R. P. Vilaplana, and W. Wertz (Eds.), *New developments in time-series analysis*. New York: Wiley & Sons.

Schmitz, B. (1987). *Zeitreihenanalyse in der Psychologie*. Weinheim, West Germany: Beltz.

Schmitz, B. (1989). *Einführung in die Zeitreihenanalyse: Modelle, Softwarebeschreibung, Anwendungen*. Bern: Huber, Switzerland.

Schwarz, G. (1978). "Estimating the dimension of a model," *Annals of Statistics*. **6,** 461–464.

Schwert, G. W. (1979). "Tests of causality: the message in the innovations." In

K. Brunner and A. H. Meltzer (Eds.), *Three aspects of policy and policymaking. Carnegie Rochester Conference Series,* Vol. 10. Amsterdam: North-Holland.

Tiao, G. C., and Box, G. E. P. (1981). "Modelling multiple time series with applications," *Journal of the American Statistical Association.* **76,** 802–816.

Tong, H. (1981). "A note on a Markov bilinear stochastic process in discrete time," *Journal of Time Series Analysis.* **2,** 279–284.

Tsay, R. S., and Tiao, G. C. (1984). "Consistent estimates of autoregressive parameters and extended sample autocorrelation function for stationary and nonstationary ARMA models," *Journal of the American Statistical Association.* **79,** 84–96.

Wiener, N. (1956). "The theory of prediction." In E. F. Beckenbach (Ed.), *Modern mathematics for engineers. Series I* (pp. 165–190). New York: McGraw-Hill.

Wilson, G. T. (1973). "The estimation of parameters in multivariate time-series models," *Journal of the Royal Statistical Society, Series B.* **35,** 76–85.

Descriptive and Associative Developmental Models

Chapter 11

JOHN TISAK

Department of Psychology
Bowling Green State University
Bowling Green, Ohio

WILLIAM MEREDITH

Department of Psychology
Univeristy of California, Berkeley
Berkeley, California

Abstract

Many of the statistical models (for example, t-tests, ANOVA, and correlational analysis) used in psychology are static, that is primarily designed for measures taken at one point in time. With increased interest in psychological development, greater emphasis has been placed on dynamic or time-dependent models, such as, time series and repeated measures ANOVA. This chapter proposes a series of models based on the following principles: (a) these models encorporate multiple populations, times, and attributes; (b) they are formulated at the individual level; (c) they are aggregable from the individual to the group; (d) they include practice effects; and (e) they may be evaluated by standard statistical criteria, such as, maximum likelihood estimation and asymptotic χ^2-tests. The techniques developed include: (a) uni- and multidimensional attributes; (b) manifest or latent variables; (c) specified or unspecified growth functions; and (d) measures of associations among the longitudinal functions. Finally, with explicit modeling equations, the concepts of longitudinal reliability and stability are defined.

387

1. Introduction

In any field of endeavor, there are always a number of approaches or focuses to the phenomena under study. In this regard, psychology has been no exception. The following two classifications of the existing psychological statistical models are possible and, in many situations, useful:

1. Static models: the measurements are made at a single time or a particular occasion.
2. Dynamic models: the measurements are made at more than one time (i.e., longitudinally).

Clearly, the former is a special case of the latter.

From a general pictorial representation of psychological data as presented in Figure 11.1, one may view the important components: time, attributes, individuals, and groups. Each of these may be isolated in the figure by a particular "slice" or plane through the appropriate rectangles. From the figure, one may visualize that most of the psychological models have been at the interindividual level and have been static. For example, they have included such common approaches as t-tests (two groups and one attribute), ANOVA (more than two groups and one attribute), correlation (one group and two attributes), regression (one group and two or more attributes), and MANOVA (more than two groups and two or more attributes).

Recently there has been greater emphasis placed on the time-dimension or dynamic models, such as time series (single group, observational unit, and attribute with multiple time points) and ANOVA with replications (multiple observations, groups and times; single attribute). Notice that each of these methods has its own restrictions (e.g., the requirement of numerous multiple time points or sphericity).

With this increased interest in time-ordered observations, there has been the very simple suggestion that the best way to study a longitudinal phenomenon is to formulate a longitudinal mathematical/statistical model of that phenomenon (Cronbach and Furby, 1970; Rogosa, Brandt, and Zimowski, 1982; Rogosa and Willett, 1985). Hence a starting place might be to state, first, what are the fundamental beliefs or axioms that such a model should entail.

1. The model should contain the capability of multiple populations, times, attributes, and individuals. That is, it should have the ability to model psychological data as presented in Figure 11.1.

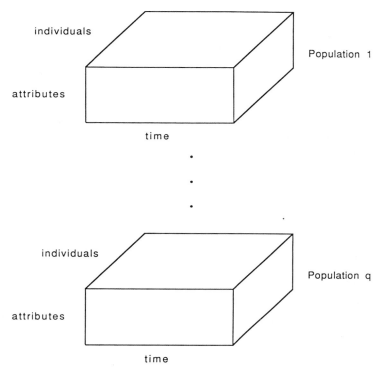

individuals

attributes

time

Population 1

individuals

attributes

time

Population q

FIGURE 11.1. A pictorial representation of psychological data.

2. The model should be focused at the individual level; that is, individuals should not be assumed unimportant. This is an idiographic or intraindividual approach.
3. The model should be aggregable to the population; that is, it should also contain the nomothetic or interindividual approach.
4. Since individuals do have a memory. The model should be able to incorporate practice effects.
5. The model must be statistically estimable and testable.

In the remainder of this chapter, an approach will be presented that conforms to these objectives. This approach is very general in that it contains either manifest or latent variables (i.e., either the observed fallible measures or the unfallible true scores). Also, it permits specified and unspecified longitudinal (developmental) curves or functions. Finally, this procedure describes the psychological data of Figure 11.1, and it allows for measures of association between the individual differences parameters of the longitudinal functions. This association may be considered analogous to a longitudinal correlation coefficient.

2. A Univariate Longitudinal Model

As mentioned in the previous section, when studying an individual longitudinally, it is important to first specify a model for development or change for that individual. Furthermore, the approach should have the ability to incorporate the group or population; that is, it should be focused at the intraindividual, but it should also contain the interindividual level. In their seminal papers, Rao (1958) and Tucker (1958) independently proposed a partial solution to this problem. They constructed a procedure that included unspecified longitudinal curves or functions.

In measuring an individual's attribute longitudinally, one may conceptualize the magnitude of that attribute as a function of characteristics of that person and time of measurement. Graphically this can be represented as in Figure 11.2.

This may be formulated more precisely and mathematically. Suppose we let $\gamma_k(t)$ be the kth unspecified longitudinal curve for all individuals and W_{ik} the weight or salience that the ith individual attaches to the $\gamma_k(t)$ curve. Then, similar to the Maclaurin or Taylor's series expansion, we may

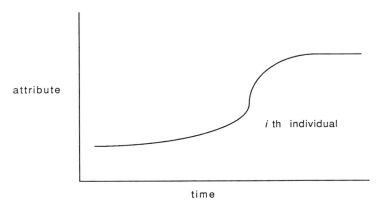

FIGURE 11.2. For a particular individual, the change in magnitude of an attribute which is a function of the characteristics of the individual and the time measurement.

express the attribute for the ith individual at time t, $Y_i(t)$, as

$$Y_i(t) = \sum_{k=1}^{d} W_{ik}\gamma_k(t) + E_i(t), \qquad (1)$$

where $i = 1, 2, \ldots, N$. $E_i(t)$ is error or residual; notice that the summation may be infinite.

The basic idea that Rao (1958) and Tucker (1958) promulgated combined the notions of lawfulness and individuality or of the nomothetic **N.B.** and idiographic approaches. That is to say, everyone develops the same way, $\gamma_k(t)$, while individual differences, W_{ik}, are ubiquitous and important. This combination of the individual and group levels of analysis is unique to the procedure. As pointed out, most of the previous psychological research has been at the group level. Or, if at the individual level, the growth functions are very explicit or specified (Guire and Kowalski, 1979). At this point, it is important to recognize that the formulation given by (1) is very general because neither the individual saliences nor the longitudinal curves are specified.

As an illustration, consider the situation where development is linear; that is, $A_i = W_{i1}$, $B_i = W_{i2}$, $\gamma_i(t) \equiv 1$, and $\gamma_2(t) = t$. Thus (1) becomes

$$Y_i(t) = A_i + B_i t + E_i(t), \qquad i = 1, 2, \ldots, N.$$

The deterministic part (i.e., $A_i + B_i t$) of this equation has the graphic representation of Figure 11.3. In this situation the individual weights, saliences, or characteristics are readily interpretable as the ith individual's initial value, A_i, and the rate of growth, B_i.

Recently, Meredith and Tisak (1990) extended the model to incorporate additional features and to permit the current standards in estimation and testing procedures. Furthermore, they recognized that certain specified functions (i.e., linearized functions) may be easily

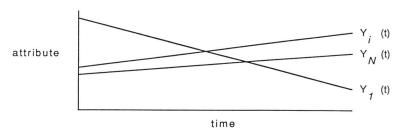

FIGURE 11.3. A plot of linear change for N individuals.

viewed as a special case of the unspecified curve approach, as clearly demonstrated by the illustration.

2.1. Matrix Reformulation

In psychological research, usually the observations take place on a small number of time points $t_1, t_2, \ldots, t_m$. The observations or measured attributes for the ith individual then becomes $Y_i(t_1), Y_i(t_2), \ldots, Y_i(t_m)$. Moreover, it is very convenient to represent (1) in matrix notation. Define the following vectors,

$$\mathbf{y}_i' = (Y_i(t_1), Y_i(t_2), \ldots, Y_i(t_m)),$$
$$\mathbf{e}_i' = (E_i(t_1), E_i(t_2), \ldots, E_i(t_m)).$$

Define the matrix

$$\Gamma = \begin{bmatrix} \gamma_1(t_1) & \gamma_2(t_1) & \cdots & \gamma_d(t_1) \\ \gamma_1(t_2) & \gamma_2(t_2) & \cdots & \gamma_d(t_2) \\ \vdots & \vdots & \vdots & \vdots \\ \gamma_1(t_m) & \gamma_2(t_m) & \cdots & \gamma_d(t_m) \end{bmatrix}.$$

Thus (1) becomes (with the i subscript dropped for notational simplicity),

$$\mathbf{y} = \Gamma\mathbf{w} + \mathbf{e}. \tag{2}$$

Assume that the average error is zero (i.e., $\mathscr{E}[\mathbf{e}] = 0$), that the errors are uncorrelated (i.e., $\mathscr{E}[\mathbf{ee}'] = \Psi$, a diagonal matrix), and that the errors are uncorrelated with the individual weights (i.e., $\mathscr{E}[\mathbf{we}'] = 0$). Also, define the first and second moments of $\mathbf{w}$ to be $\mathscr{E}[\mathbf{w}] = \tau$, and $\mathscr{E}[\mathbf{ww}'] = \Phi$, respectively. $\mathscr{E}[\cdot]$ is the expectation operator. Then

$$\mathscr{E}[\mathbf{y}] = \eta = \Gamma\tau \tag{3}$$

and

$$\mathscr{E}[\mathbf{yy}'] = \Xi = \Gamma\Phi\Gamma' + \Psi. \tag{4}$$

One recognizes (2), (3), and (4) as the basic equations of factor analysis with means (Harman, 1976; Mulaik, 1972). Notice that the factor pattern matrix Γ contains the growth curves. The realizations or estimates of $\mathbf{w}$ are the factor scores or, specific to our situation, the individual saliences.

Using super or partitioned matrices, we can write (3) and (4) compactly as

$$\mathbf{M} = \begin{bmatrix} \Xi & \eta \\ \eta' & 1 \end{bmatrix} = \begin{bmatrix} \Gamma & 0 \\ 0' & 1 \end{bmatrix} \begin{bmatrix} \Phi & \tau \\ \tau' & 1 \end{bmatrix} \begin{bmatrix} \Gamma' & 0 \\ 0' & 1 \end{bmatrix} + \begin{bmatrix} \Psi & 0 \\ 0' & 0 \end{bmatrix}. \tag{5}$$

By using confirmatory structural equation modeling packages, such as LISREL VI (Jöreskog and Sörbom, 1984), EQS (Bentler, 1989), or COSAN (McDonald, 1978), one has available standard statistical techniques, such as estimation of the parameters (Γ, Φ, τ, and Ψ) and asymptotic χ^2-tests of the goodness-of-fit of the model.

2.2. Longitudinal Reliability

Although the concept of static reliability (i.e., reliability of a measure at a single time) has been investigated thoroughly (Lord and Novick, 1968), a formulation of a dynamic or longitudinal reliability has received little attention. In the former case, one way of defining reliability for a measure is the ratio of its true score variance to its error score variance. Extending this definition, we propose a longitudinal definition of reliability in this section.

First, notice that the covariance matrices of $\mathbf{y}$ and $\mathbf{w}$ are given, respectively, by

$$\Sigma_{yy'} = \Xi - \eta\eta'$$

and

$$\Sigma_{ww'} = \Phi - \tau\tau'.$$

Next, with the longitudinal curve model, the covariance matrix of the observations, $\mathbf{y}$, is

$$\Sigma_{yy'} = \Gamma\Sigma_{ww'}\Gamma' + \Psi.$$

Let $D_{\Gamma\Sigma_{ww'}\Gamma'}$ be a diagonal matrix with its nonzero elements from the main diagonal of $\Gamma\Sigma_{ww'}\Gamma$ and let $\mathbf{1}_m$ be a unit vector of order m.

Consistent with the definition of reliability given above, longitudinal reliability, ρ_{yy}, may be given as

$$\rho_{yy} = D_{\Gamma\Sigma_{ww'}\Gamma'}[D_{\Gamma\Sigma_{ww'}\Gamma'+\Psi}]^{-1}\mathbf{1}_m.$$

Observe that ρ_{yy} is no longer a scalar, but it is instead an $m \times 1$ vector. With this more general formulation of reliability, one has, within the framework of univariate (and subsequently multivariate) longitudinal models, a way of assessing changes in reliability.

2.3. Stability

The stability of an attribute has been an interesting and important issue (Bloom, 1964; Rogosa, Floden, and Willett, 1984). Also Wohlwill (1973) has produced very precise categories for different types of stability. This section will be limited to particular parameterizations of (1), which will yield specific types of stability.

One may say that *strict stability* exists when $d = 1$ and $\gamma_1(t) \equiv 1$. In this case,

$$Y_i(t) = W_{i1} + E_i(t);$$

that is, individuals do not change across time.

If $d = 2$, $\gamma_1(t) \equiv 1$, $\gamma_2(t) = t$, and $W_{i2} = W_2$, then there is *parallel stability*; that is,

$$Y_i(t) = W_{i1} + W_2 t + E_i(t).$$

In this situation individuals develop linearly, but individual differences in an attribute do not vary across time.

Let W_{i1} be directly proportional to W_{i2} or to be equal to a constant, $\gamma_1(t) \equiv 1$, $\gamma_2(t) = t$, and $d = 2$. One may define this situation as *linear stability*; that is,

$$Y_i(t) = W_{i1} + W_{i2} t + E_i(t).$$

For this case, the differences among individuals will vary longitudinally.

Of practical importance is *monotonic stability*. In this condition, individuals maintain their same rank order throughout the measured time periods. Hence,

$$Y_i(t) = W_{i1} + W_{i2}\gamma_1(t) + E_i(t),$$

where $d = 2$, $\gamma_1(t) \equiv 1$, $\gamma_2(t)$ is any monotonic function, and W_{i1} is directly proportional to W_{i2} or equal to 0.

The stability conditions (strict, parallel, linear, and monotonic) are each in turn less restricted (i.e., the situations are nested). By the methods proposed in this chapter, each of these types of stability is directly estimable and testable.

2.4. An Extension for Period or Practice Effects

At the onset of this chapter, we recognized that individuals are influenced by previous performance, and thus there is the possibility of carryover effects. These effects may be incorporated in the model by additive and/or multiplicative parameters (Vinsonhaler and Meredith, 1966).

Equation (1) thus becomes

$$Y_i(t) = \delta(t)\left[\sum_{k=1}^{d} W_{ik}\gamma_k(t) + \alpha(t)\right] + E_i(t), \tag{6}$$

where the additive and multiplicative period/practice effects are given by $\alpha(t)$ and $\delta(t)$, respectively. For discrete time, these effects in (6) may compactly be incorporated into (2) and (5) as

$$\mathbf{y} = D_\delta[\mathbf{\Gamma w} + \boldsymbol{\alpha}] + \mathbf{e} \tag{7}$$

and

$$\mathbf{M} = \begin{bmatrix} D_\delta & \mathbf{0} \\ \mathbf{1}' & 1 \end{bmatrix}\begin{bmatrix} \mathbf{\Gamma} & \boldsymbol{\alpha} \\ \mathbf{0}' & 1 \end{bmatrix}\begin{bmatrix} \mathbf{\Phi} & \boldsymbol{\tau} \\ \boldsymbol{\tau}' & 1 \end{bmatrix}\begin{bmatrix} \mathbf{\Gamma}' & \mathbf{0} \\ \boldsymbol{\alpha}' & 0 \end{bmatrix}\begin{bmatrix} D_\delta & \mathbf{0} \\ \mathbf{0}' & 1 \end{bmatrix} + \begin{bmatrix} \mathbf{\Psi} & \mathbf{0} \\ \mathbf{0}' & 0 \end{bmatrix}, \tag{8}$$

where D_δ is a diagonal matrix with diagonal elements $\{\delta(t_1), \delta(t_2), \ldots, \delta(t_m)\}$ and $\boldsymbol{\alpha}' = (\alpha(t_1), \alpha(t_2), \ldots, \alpha(t_m))$.

2.5. An Extension for Multiple Populations

Another useful feature for a longitudinal model is the capability to deal with multiple populations or cohorts. Let the cth population be indicated by the superscript c, where $c \in \{1, 2, \ldots, q\}$. Then (7) and (8), which include practice effects, may be rewritten as

$$\mathbf{y}^{(c)} = D_\delta^{(c)}[\mathbf{\Gamma}^{(c)}\mathbf{w}^{(c)} + \boldsymbol{\alpha}^{(c)}] + \mathbf{e}^{(c)}$$

and

$$M^{(c)} = \begin{bmatrix} D_\delta^{(c)} & \mathbf{0} \\ \mathbf{0}' & 1 \end{bmatrix}\begin{bmatrix} \mathbf{\Gamma}^{(c)} & \boldsymbol{\alpha}^{(c)} \\ \mathbf{0}' & 1 \end{bmatrix}\begin{bmatrix} \mathbf{\Phi}^{(c)} & \boldsymbol{\tau}^{(c)} \\ \boldsymbol{\tau}^{(c)'} & 1 \end{bmatrix}\begin{bmatrix} \mathbf{\Gamma}^{(c)'} & \mathbf{0} \\ \boldsymbol{\alpha}^{(c)'} & 1 \end{bmatrix}\begin{bmatrix} D_\delta^{(c)} & \mathbf{0} \\ \mathbf{0}' & 1 \end{bmatrix} + \begin{bmatrix} \mathbf{\Psi}^{(c)} & \mathbf{0} \\ \mathbf{0}' & 0 \end{bmatrix}.$$

Often it is convenient and reasonable to assume that the period/practice effects are invariant across populations: $D_\delta^{(c)} = D_\delta$ and $\alpha^{(c)} = \alpha$. Of special interest is when the populations are defined by different birth year cohorts with all cohorts measured at the same time points. Such longitudinal designs are usually referred to as *cohort sequential designs* (Schaie, 1965).

As an example of this situation, if the kth column of each of the $\mathbf{\Gamma}^{(c)}$ is plotted as a function of time, for four cohorts, the graph might look like Figure 11.4.

Situations where the longitudinal functions for these cohort sequential designs can be overlapping are especially important. For this case the growth curve can be constructed to be equal in a lagged fashion. More

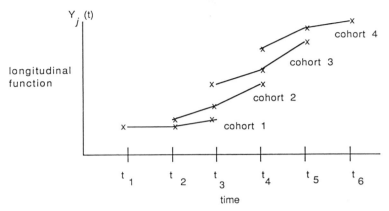

FIGURE 11.4.   Curves for a cohort sequential design.

precisely for the cohorts presented in the example, constrain

$$
\begin{array}{llll}
\gamma_{jt_1} & \gamma_j^{(1)}(t_1) & & \\
\gamma_{jt_2} & \gamma_j^{(1)}(t_2) = \gamma_j^{(2)}(t_1) & & \\
\gamma_{jt_3} & \gamma_j^{(1)}(t_3) = \gamma_j^{(2)}(t_2) & = \gamma_j^{(3)}(t_1) & \\
\gamma_{jt_4} & \qquad\quad\;\; \gamma_j^{(2)}(t_3) & = \gamma_j^{(3)}(t_2) & = \gamma_j^{(4)}(t_1) \\
\gamma_{jt_5} & & \gamma_j^{(3)}(t_3) & = \gamma_j^{(4)}(t_2) \\
\gamma_{jt_6} & & & \gamma_j^{(4)}(t_3)
\end{array}
$$

where the new curve $(\gamma_{jt_1}, \gamma_{jt_2}, \gamma_{jt_3}, \gamma_{jt_4}, \gamma_{jt_5}, \gamma_{jt_6})$ holds for all individuals in all cohorts. This powerful technique allows one to measure several different birth year cohorts for shorter periods of time and then to extend the function for the entire period spanned by all ages.

3. A Multivariate Longitudinal Model

The multivariate generalization of the growth curve model was originally proposed by Tucker (1966) as a data descriptive technique. Subsequently, Tisak and Meredith (June 1986) reformulated the approach as a statistical model and extended the previous univariate longitudinal model. This approach will be introduced in this section. However, in keeping with the complexities of the multivariate situation, the notational scheme from the previous section must be extended and, to some extent, redefined.

3.1. The Longitudinal Factor-Analytic Model in Multiple Populations.

Let the variables be indexed by $j \in \{1, 2, \ldots, p\}$. Let the different populations or cohorts be indicated by $c \in \{1, 2, \ldots, q\}$. Let m_c be the number of measurement periods for the cth cohort. Let $s, t \in \{1, 2, \ldots, m_c\}$ index the occasion of measurement for a given cohort. Let the factors be indexed by $g, h \in \{1, 2, \ldots, r\}$. Next define the random vectors. $\mathbf{x}_s^{(c)}$ is the $p \times 1$ manifest or observed variable vector in cohort c on occasion s; analogously, $\mathbf{z}_s^{(c)}$ is the $r \times 1$ common factor vector. $\mathbf{u}_s^{(c)}$ is the corresponding $p \times 1$ unique factor vector. As before, the subscript i has been dropped for convenience. Define $\boldsymbol{\Lambda}$ to be a $p \times r$ factor pattern or regression weight matrix.

A factor analytic model may be written as

$$\mathbf{x}_s^{(c)} = \boldsymbol{\Lambda} \mathbf{z}_s^{(c)} + \mathbf{u}_s^{(c)},$$

where $\mathscr{E}[\mathbf{x}_s^{(c)}] = \boldsymbol{\mu}_s^{(c)}$, $\mathscr{E}[\mathbf{z}_s^{(c)}] = \mathbf{v}_s^{(c)}$, and $\mathscr{E}[\mathbf{u}_s^{(c)}] = \mathbf{0}$. Also define the matrices $\mathscr{E}[\mathbf{x}_s^{(c)}, \mathbf{x}_t^{(c)\prime}] = \boldsymbol{\Omega}_{st}^{(c)}$, $\mathscr{E}[\mathbf{z}_s^{(c)}, \mathbf{z}_t^{(c)\prime}] = \mathbf{T}_{st}^{(c)}$, and $\mathscr{E}[\mathbf{u}_s^{(c)}, \mathbf{u}_t^{(c)\prime}] = \boldsymbol{\Psi}_{st}^{(c)}$, a diagonal matrix. Then

$$\mathscr{E}[\mathbf{x}_s^{(c)}] = \boldsymbol{\Lambda} \mathbf{v}_s^{(c)}$$

and

$$\mathscr{E}[\mathbf{x}_s^{(c)} \mathbf{x}_t^{(c)\prime}] = \boldsymbol{\Lambda} \mathbf{T}_{st}^{(c)} \boldsymbol{\Lambda}' + \boldsymbol{\Psi}_{st}^{(c)}.$$

In this model, there are the implicit assumptions that the unique factors are uncorrelated with the common factors both within and across occasions. Furthermore, there are also the implicit assumptions that the unique factors are uncorrelated with themselves within occasion, but may be correlated with the same unique factor across occasions. Finally, there is the assumption that the factor pattern matrix remains invariant across both occasions and cohorts. For more relaxed conditions on the factor pattern matrix refer to Tisak and Meredith (1989).

Let $\mathbf{x}^{(c)}$, $\mathbf{z}^{(c)}$, $\mathbf{u}^{(c)}$, $\boldsymbol{\mu}^{(c)}$, $\mathbf{v}^{(c)}$, $\boldsymbol{\Omega}^{(c)}$, $\mathbf{T}^{(c)}$, and $\boldsymbol{\Psi}^{(c)}$ be partitioned vectors and matrices with partitioned elements $\mathbf{x}_s^{(c)}$, $\mathbf{z}_s^{(c)}$, $\mathbf{u}_s^{(c)}$, $\boldsymbol{\mu}_s^{(c)}$, $\mathbf{v}_s^{(c)}$, $\boldsymbol{\Omega}_{st}^{(c)}$, $\mathbf{T}_{st}^{(c)}$, and $\boldsymbol{\Psi}_{st}^{(c)}$, respectively. Because of its structure $\boldsymbol{\Psi}^{(c)}$ will be referred to as a *band diagonal* matrix. In order for the factors to have the same meaning across both time (stationarity) and population (invariance), notice that $D_{\boldsymbol{\Lambda}}$ is an *$mp \times mr$ block diagonal* matrix:

$$D_{\boldsymbol{\Lambda}} = \begin{bmatrix} \boldsymbol{\Lambda} & 0 & \cdots & 0 \\ 0 & \boldsymbol{\Lambda} & \cdots & 0 \\ \vdots & \vdots & \vdots & \vdots \\ 0 & 0 & \cdots & \boldsymbol{\Lambda} \end{bmatrix}.$$

Utilizing these partitioned matrices, we have equations for longitudinal factor analysis in several populations:

$$\mathbf{x}^{(c)} = D_\Lambda \mathbf{z}^{(c)} + \mathbf{u}^{(c)}, \tag{9}$$

and

$$\boldsymbol{\mu}^{(c)} = D_\Lambda \mathbf{v}^{(c)}, \tag{10}$$

$$\boldsymbol{\Omega}^{(c)} = D_\Lambda T^{(c)} D_\Lambda' + \boldsymbol{\Psi}^{(c)}. \tag{11}$$

Often, however, a reformulation of (9), (10), and (11) is more convenient both for mathematical derivations and for detecting numerical patterns in the data matrices. Let $P_1^{(c)}$ and $P_2^{(c)}$ be permutation matrices that order the variables and factors, respectively, by time of measurement. To illustrate this, let $\mathbf{x}_s^{(c)'} = [X_{s1}^{(c)}, X_{s2}^{(c)}, \ldots, X_{sp}^{(c)}]$, where $X_{sj}^{(c)}$ is a scalar random variable for the ith variable on the sth occasion. The original random variable vector may be represented as

$$\mathbf{x}^{(c)'} = [\mathbf{x}_1^{(c)'}, \mathbf{x}_2^{(c)'}, \ldots, \mathbf{x}_m^{(c)'}]$$
$$= [X_{11}^{(c)}, X_{12}^{(c)}, \ldots, X_{1p}^{(c)}, \ldots, X_{m1}^{(c)}, X_{m2}^{(c)}, \ldots, X_{mp}^{(c)}].$$

Permuted this becomes

$$(P_1^{(c)} \mathbf{x}^{(c)})' = \mathbf{x}^{(c)'} P_1^{(c)'} = [X_{11}^{(c)}, x_{21}^{(c)}, \ldots, X_{m1}^{(c)}, \ldots, X_{1p}^{(c)}, X_{2p}^{(c)}, \ldots, X_{mp}^{(c)}].$$

The original variables are grouped within the time of measurement, whereas the permuted variables are grouped within a particular variable for all time periods. The permutations of the factors are analogous. Define

$$\mathbf{y}^{(c)} = P_1^{(c)} \mathbf{x}^{(c)},$$

$$\mathbf{f}^{(c)} = P_2^{(c)} \mathbf{z}^{(c)},$$

$$\mathbf{v}^{(c)} = P_1^{(c)} \mathbf{u}^{(c)},$$

$$\boldsymbol{\eta}^{(c)} = P_1^{(c)} \boldsymbol{\mu}^{(c)},$$

$$\boldsymbol{\xi}^{(c)} = P_2^{(c)} \mathbf{v}^{(c)},$$

$$\boldsymbol{\Xi}^{(c)} = P_1^{(c)} \boldsymbol{\Omega}^{(c)} P_1^{(c)'},$$

$$\boldsymbol{\Phi}^{(c)} = P_2^{(c)} T^{(c)} P_2^{(c)'},$$

$$\boldsymbol{\Upsilon}^{(c)} = P_1^{(c)} \boldsymbol{\Psi}^{(c)} P_1^{(c)'},$$

and

$$\boldsymbol{\Delta} = P_1^{(c)} D_\Lambda P_2^{(c)'}.$$

Notice that $\boldsymbol{\Delta}$ is a quasi-band diagonal matrix and that $\boldsymbol{\Upsilon}^{(c)}$ is a block diagonal matrix with symmetrical blocks ($\Upsilon_{jj}^{(c)} = \Upsilon_{jj}^{(c)'}$); that is,

$$\Upsilon^{(c)} = \begin{bmatrix} \Upsilon_1^{(c)} & 0 & \cdots & 0 \\ 0 & \Upsilon_2^{(c)} & \cdots & 0 \\ \vdots & \vdots & \vdots & \vdots \\ 0 & 0 & \cdots & \Upsilon_p^{(c)} \end{bmatrix}$$

and

$$\Delta = \begin{bmatrix} \Delta_{11} & \Delta_{12} & \cdots & \Delta_{1r} \\ \Delta_{21} & \Delta_{22} & \cdots & \Delta_{2r} \\ \vdots & \vdots & \vdots & \vdots \\ \Delta_{p1} & \Delta_{p2} & \cdots & \Delta_{pr} \end{bmatrix},$$

where $\Delta_{sg} = \lambda_{sg}I$, is a diagonal matrix for factor g on occasion s. With these permutations (9), (10), and (11) may be rewritten as

$$y^{(c)} = \Delta f^{(c)} + v^{(c)}, \tag{12}$$

$$\eta^{(c)} = \Delta \xi^{(c)}, \tag{13}$$

and

$$\Xi^{(c)} = \Delta \Theta^{(c)} \Delta' + \Upsilon^{(c)}. \tag{14}$$

Equations (12), (13), and (14) are then a restatement of the standard factor analytic model with means; only the variables have been reordered to facilitate the expression of the longitudinal curve model in the next section.

3.2. A Latent Variable Longitudinal Curve Model

Similar to the formulation of (1), assume that $f_g^{(c)}$ can be decomposed as a sum of elementary (generally unknown) "basis" functions. That is,

$$f_g^{(c)} = \Gamma_g^{(c)} w_g^{(c)} + s_g^{(c)}, \tag{15}$$

where $f_g^{(c)}$ is an $m \times 1$ vector for the gth factor and the cth cohort; its elements contain the values for the occasions of measurement. The $m \times d$ matrix $\Gamma_g^{(c)}$ contains the basis functions. Usually, for $\Gamma_g^{(c)}$ to be meaningful, additional constraints must be imposed across cohorts (e.g., "lagged" equality of elements). The random vector $w_g^{(c)}$ contains the weights or saliences that characterize the individual. They present the degree to which an individual "partakes" of the basis functions $\Gamma_g^{(c)}$. The errors are given in the random vector $s_g^{(c)}$. Finally, let $\mathscr{E}[\mathbf{f}_g^{(c)}] = \boldsymbol{\Gamma}_g^{(c)}\boldsymbol{\tau}_g^{(c)}$ and $\mathscr{E}[\mathbf{f}_g^{(c)}\mathbf{f}_h^{(c)'}] = \boldsymbol{\Gamma}_g^{(c)}\boldsymbol{\Phi}_{gh}^{(c)}\boldsymbol{\Gamma}_h^{(c)'} + D_{gh}^{(c)}$, where $D_{gg}^{(c)}$ is a diagonal matrix and $D_{gh}^{(c)} = \mathbf{0}$ if $g \neq h$. More complex structures also may be imposed on $D_{gh}^{(c)}$.

3.3. The Longitudinal Factor-Analytic Model in Multiple Populations with Longitudinal Curves

With the previous definitions, one is now ready to consider the multivariate model. Let $\mathbf{f}^{(c)}$, $\mathbf{w}^{(c)}$, $\mathbf{s}^{(c)}$, $\boldsymbol{\tau}^{(c)}$, $\boldsymbol{\Phi}^{(c)}$, and $D^{(c)}$ be partitioned vectors and matrices with partitioned elements $\mathbf{f}_g^{(c)}$, $\mathbf{w}_g^{(c)}$, $\mathbf{s}_g^{(c)}$, $\boldsymbol{\tau}_g^{(c)}$, $\boldsymbol{\Phi}_{gh}^{(c)}$, and $D_{gh}^{(c)}$, respectively. Notice that $\boldsymbol{\Gamma}^{(c)}$ is an $mr \times rd$ block diagonal matrix:

$$\boldsymbol{\Gamma}^{(c)} = \begin{bmatrix} \boldsymbol{\Gamma}_1^{(c)} & \mathbf{0} & \cdots & \mathbf{0} \\ \mathbf{0} & \boldsymbol{\Gamma}_2^{(c)} & \cdots & \mathbf{0} \\ \vdots & \vdots & \vdots & \vdots \\ \mathbf{0} & \mathbf{0} & \cdots & \boldsymbol{\Gamma}_r^{(c)} \end{bmatrix}.$$

By substituting these equations into (12)–(15), we may formulate the *fundamental equation for longitudinal factor analysis in multiple populations with developmental curves* as

$$\mathbf{y}^{(c)} = \boldsymbol{\Delta}\mathbf{f}^{(c)} + \mathbf{v}^{(c)} = \boldsymbol{\Delta}[\boldsymbol{\Gamma}^{(c)}\mathbf{w}^{(c)} + \mathbf{s}^{(c)}] + \mathbf{v}^{(c)}, \tag{16}$$

$$\boldsymbol{\eta}^{(c)} = \boldsymbol{\Delta}\boldsymbol{\xi}^{(c)} = \boldsymbol{\Delta}\boldsymbol{\Gamma}^{(c)}\boldsymbol{\tau}^{(c)} \tag{17}$$

and

$$\boldsymbol{\Xi}^{(c)} = \boldsymbol{\Delta}\boldsymbol{\Theta}^{(c)}\boldsymbol{\Delta}' + \mathbf{Y}^{(c)} = \boldsymbol{\Delta}[\boldsymbol{\Gamma}^{(c)}\boldsymbol{\Phi}^{(c)}\boldsymbol{\Gamma}^{(c)'} + D^{(c)}]\boldsymbol{\Delta}' + \mathbf{Y}^{(c)}. \tag{18}$$

Notice that as stated, (16), (17), and (18) are simply an analysis of covariance structures (ACOVS) model for several populations (Jöreskog, 1970, 1971) with means. Furthermore, a solution to (18) may be obtained by any maximum likelihood confirmatory factor analysis program, such as LISREL (Jöreskog and Sörbom, 1984), EQS (Bentler, 1989), or COSAN (Mcdonald, 1978).

3.2. An Extension for Period or Practice Effects

As before, the longitudinal factor analysis model in multiple populations with developmental curves may be extended to allow for additive period/practice effects on either the manifest or latent variables. To incorporate additive practice effects on the manifest variables, (16) becomes

$$\mathbf{y}^{(c)} = \boldsymbol{\alpha} + \boldsymbol{\Delta}\mathbf{f}^{(c)} + \mathbf{v}^{(c)} = \boldsymbol{\alpha} + \boldsymbol{\Delta}[\boldsymbol{\Gamma}^{(c)}\mathbf{w}^{(c)} + \mathbf{s}^{(c)}] + \mathbf{v}^{(c)}. \tag{19}$$

Allowing for these effects on the latent variables, (15) when written as a supermatrix becomes

$$\mathbf{f}^{(c)} = \boldsymbol{\kappa} + \boldsymbol{\Gamma}_g^{(c)}\mathbf{w}_g^{(c)} + \mathbf{s}_g^{(c)}. \tag{20}$$

For both (19) and (20) identifiability constraints must be imposed; for example, the effect corresponding to each variable or factor at first measurement is set equal to zero. Note that these effects may best be assessed when there are multiple populations.

3.5. A General Approach with a Number of Special Cases

As formulated in this section, the model is a quite general approach to individual development. Besides the desired assets described in the introduction, it contains the following special cases:

1. The univariate longitudinal model as presented in the second section is a special situation. This is where only one variable is repeatedly measured.
2. The developmental curves may be specified or unspecified. If a known function can be linearized, then the procedure is exact. For example, the columns of $\boldsymbol{\Gamma}^{(c)}$ may contain orthogonal polynomial coefficients of known degree. On the other hand, if the known function cannot be linearized, then the procedure is an approximation. For the Rao-Tucker situation where the function is unknown, $\boldsymbol{\Gamma}^{(c)}$ may be left unspecified.
3. The curves may be spline functions, where the appropriate coefficients are placed into $\boldsymbol{\Gamma}^{(c)}$. Splines are defined to be piecewise polynomial, where each polynomial segment is usually of low degree (e.g., either quadratic or cubic). Although approaches to splines are complicated (cf. Schumaker, 1981), in the present situation the use of truncated polynomials or "+" functions may be conveniently implemented (i.e., estimated and tested). This procedure is discussed and illustrated in Smith (1979).
4. For the univariate and multivariate models, the longitudinal functions may be directly equated to the manifest variables.
5. For the multivariate model, the longitudinal curves may either be equated to the manifest variable or to a latent variable.

4. Descriptive Developmental Models

One aspect of the multivariate longitudinal model presented in a previous section is that it *describes* the way that an individual develops. This is a first step of any longitudinal methodology, because it is important to know whether an attribute increases, decreases or remains constant as a function of time. At the group or interindividual level, often this change is evident from the plots of the longitudinal functions (i.e., the columns of $\Gamma_g^{(c)}$). These graphs allow one to deduce the lawfulness of the phenomena. For the individual or at the intraindividual level, individual saliences may be estimated. Furthermore, the individual's predicted developmental curve may be plotted by using the deterministic terms of (1) or by using the deterministic terms of an extension to (1).

5. Associative Developmental Models

Another important facet of the multivariate longitudinal model is that this model gives associations among the individual differences parameters. These associations are analogous to the static model's correlation coefficient. In fact, they are crucial to any scientific study of development because they indicate the influences of development or, to put it another way, they are correlates of change.

More specifically from (17) and (18), the first and second moments of $\mathbf{w}^{(c)}$ are given by $\tau^{(c)}$ and $\Phi^{(c)}$, respectively. Thus the covariance matrix of $\mathbf{w}^{(c)}$ is given by

$$\Sigma_{\mathbf{w}^{(c)}\mathbf{w}^{(c)'}} = \Phi^{(c)} - \tau^{(c)}\tau^{(c)'},$$

and the correlation matrix of $\mathbf{w}^{(c)}$ is obtained by

$$R_{\mathbf{w}^{(c)}\mathbf{w}^{(c)'}} = D_{\mathbf{w}^{(c)}\mathbf{w}^{(c)'}}^{-1/2} \Sigma_{\mathbf{w}^{(c)}\mathbf{w}^{(c)'}} D_{\mathbf{w}^{(c)}\mathbf{w}^{(c)'}}^{-1/2}.$$

Again recall that $\mathbf{w}^{(c)'} = (\mathbf{w}_1^{(c)'}, \mathbf{w}_2^{(c)'}, \ldots, \mathbf{w}_r^{(c)'})$; that is, $\mathbf{w}_g^{(c)}$ contains the individual weight that describes (parameterizes) the gth latent variable. Hence, one may see the association between an individual salience from one function and an individual salience from another function. Notice that Rogosa and Willett's (1985) discussion on correlates of change is a special case of this presentation. In their presentation, one of the variates is assumed static, whereas the other is considered dynamic.

6. A Simple Example

This illustration uses longitudinal data[1] on sociological and psychological variates collected at three time points (1966, 1967, 1971). These data ($N = 932$) were reported in an example (Wheaton et al., 1977). A more complete description is given in the original paper by Summers et al., (1969). In particular, Bogardus's (1925) Latin American social distance scale, which comprises a summated 6 item scale with each item coded by a dichotomous acceptance-rejection designation and a measure of educational attainment, was analyzed. See Table 11.1.

6.1. A Descriptive Developmental Model

As an example of the univariate longitudinal model, these data were described as follows. Since education was measured only at 1966, the model for $Y_{i0}(t)$ is completely predetermined from (1) as

$$Y_{i0}(t) \equiv W_{i0}.$$

On the other hand, the Latin American social distance scale, $Y_{i1}(t)$, is dynamic, and from (1) may be represented by

$$Y_{i1}(t) = \gamma_1(t)W_{i1} + E_i(t). \tag{21}$$

Its mean, variance and covariances are given, respectively, by

$$\mathscr{E}[Y_{i1}(t)] = \eta(t) = \gamma_1(t)\tau_1,$$
$$\sigma^2_{Y_1(t)} = \gamma_1(t)^2\sigma^2_{W_1} + \sigma^2_E$$

TABLE 11.1. Correlation Matrix of These Variates

		Y_0	Y_{11}	T_{12}	Y_{13}
Education (1966)	Y_0	1.00			
Latin American social distance (1966)	Y_{11}	−.25	1.00		
Latin American social distance (1967)	Y_{12}	−.16	.49	1.00	
Latin American social distance (1971)	Y_{13}	−.22	.47	.41	1.00
	Mean	10.90	1.11	.89	.78
	S.D.	3.10	1.40	1.27	1.25

[1] The authors wish to thank Gene Summers and John Scott for permitting us to use these data.

and

$$\sigma_{Y_1(t)Y_1(t')} = \gamma_1(t)\gamma_1(t')\sigma_{W_1}^2, \qquad \text{for } t \neq t'.$$

In this example, $t, t' \in \{1966, 1967, 1971\}$, and the errors are assumed to be homoscedastic (i.e., $\sigma_{E(t)}^2 = \sigma_E^2$). Furthermore, $\gamma_1(t)$ is the unspecified developmental or change function.

The social distance model was analyzed in LISREL VI (Jöreskog and Sörbom, 1984) to yield the following results. The fit of the model was assessed by the significance of the χ^2-statistic: $\chi^2(4) = 2.33$, $p = .675$. Given the sample size, this model fits the data exceedingly well. Parenthetically, other models, such as a static one ($\gamma_1(t) \equiv 1$), significantly did not fit these data.

Maximum likelihood estimates were also obtained: $\hat{\gamma}(1) \equiv 1$, $\hat{\gamma}(2) = .81$, $\hat{\gamma}(3) = .73$, $\hat{\tau}_1 = 1.10$, $\hat{\sigma}_{W_1}^2 = 1.09$, and $\hat{\sigma}_E^2 = .92$. In particular, notice that social distance to Latin Americans monotonically decreased from 1966 to 1971. From (21), the measured attribute may be considered monotonically stable. That is, individuals do not change in their rank ordering across time. Reliabilities may be calculated by

$$\hat{\rho}(t) \equiv \hat{\rho}_{X(t)X(t)} = \frac{\hat{\gamma}(t)^2\hat{\sigma}_{W_1}^2}{\hat{\gamma}(t)^2\hat{\sigma}_{W_1}^2 + \hat{\sigma}_E^2}.$$

The following estimates were obtained: $\hat{\rho}(1) = .54$, $\hat{\rho}(2) = .49$, and $\hat{\rho}(3) = .39$.

7. An Associative Model

Given this descriptive model of development, one may ask what are the correlates of change. In particular, what is the correlation between educational attainment, Y_0, and the "rate" of change in the Latin American social distance scale, W_1 (i.e., $\rho_{Y_0W_1}$). A LISREL VI analysis yielded $\hat{\rho}_{Y_0W_1} = -.31$. The fit of this model was $\chi^2(6) = 7.79$, $p = .254$. From this result, one may conclude that educational attainment is negatively related to the "rate" of decrease in social distance (and, as a consequence, to social distance longitudinally).

References

Bentler, P. M. (1989). EQS *structural equations program manual*. Los Angeles: BMDP Statistical Software, Inc.

Bloom, B. S. (1964). *Stability and change in human characteristics* (pp. 95–131). New York: Wiley.

Bogardus, E. S. (1925). "Measuring social distances," *Journal of Applied Sociology*. **9**, 299–308.

Cronbach, L. J., and Furby, L. (1970). "How should we measure "change"—or should we?" *Psychological Bulletin*. **74**, 68–80.

Guire, K. E., and Kowalski, C. J. (1979). "Mathematical description and representation of developmental change functions on the intra- and interindividual levels." In J. R. Nesselroade and P. B. Baltes (Eds.), *Longitudinal research in the study of behavior and development* (pp. 89–110). New York: Academic Press.

Harman, H. H. (1976). *Modern factor analysis*. Chicago: University of Chicago Press.

Jöreskog, K. G. (1970). "A general method for analysis of covariance structures," *Biometrika*. **37**, 239–251.

Jöreskog, K. G. (1971). "Simultaneous factor analysis in several populations," *Psychometrika*. **36**, 409–426.

Jöreskog, K. G., and Sörbom, D. (1984). *LISREL VI: Analysis of linear structural relationships by the method of maximum likelihood*. Chicago: Scientific Software.

Lord, F. M., and Novick, M. R. (1968). *Statistical theories of mental test scores*. Menlo Park, Calif.: Addison-Wesley.

McDonald, R. P. (1978). "A simple comprehensive model for the analysis of covariance structures," *British Journal of Mathematical and Statistical Psychology*. **31**, 59–72.

Meredith, W., and Tisak, J. (1984, June). "Tuckerizing curves." Paper presented at the Annual Meeting of the Psychometric Society, Santa Barbara, Calif.

✓Meredith, W., and Tisak, J. (1990). "Latent curve analysis." *Psychometrika*, **55**.

Mulaik, S. A. (1972). *The foundation of factor analysis*. New York: McGraw-Hill.

Rao, C. R. (1958). "Some statistical methods for comparison of growth curves," *Biometrics*. **14**, 1–17.

Rogosa, D. R., Brandt, D., and Zimowski, M. (1982). "A growth curve approach to the measurement of change," *Psychological Bulletin*. **90**, 726–748.

Rogosa, D. R., Floden, R. E., and Willett, J. B. (1984). "Assessing the stability of teacher behavior," *Journal of Educational Psychology*. **76**, 1000–1028.

Rogosa, D. R., and Willitt, J. B. (1985). "Understanding correlates of change by modeling individual differences in growth," *Psychometrika*. **50**, 203–228.

Schaie, K. W. (1965). "A general model for the study of developmental problems," *Psychological Bulletin*. **64**, 92–107.

Schumaker, L. L. (1981). *Spline functions: basic theory*. New York: Wiley.

Smith, P. L. (1979). "Splines as a useful and convenient statistical tool," *American Statistician*. **33**, 57–62.

Summers, G. F., Hough, R. L., Scott J. T., and Folse, C. L. (1969). *Before industralization: a rural social system base study.* Bulletin No. 736. Urbana, Ill.: Illinois Agricultural Experiment Station. University of Illinois.

Tisak, J., and Meredith, W. (1986, June). "Tuckerizing' curves for latent variables." Paper presented at the annual meeting of the Psychometric Society, Toronto, Canada.

Tisak, J., and Meredith, W. (1989). "Exploratory longitudinal factor analysis in multiple populations," *Psychometrika.* **54,** 261–281.

Tucker, L. R. (1958). "Determination of parameters of a functional relation by factor analysis," *Psychometrika.* **23,** 19–23.

Tucker, L. R. (1966). "Learning theory and multivariate experiment: illustration of generalization learning curves." In R. B. Cattell (Ed.), *Handbook of multivariate experimental psychology* (pp. 476–501). Chicago: Rand McNally.

Vinsonhaler, J. F., and Meredith, W. (1966). "A stochastic model for repeated testing," *Multivariate Behavioral Research.* **1,** 461–477.

Wheaton, B., Muthén, B., Alwin, D. F., and Summers, G. F. (1977). "Assessing reliability and stability in panel models." In D. R. Heise (Ed.), *Sociological methodology 1977* (pp. 84–136). San Francisco: Jossey-Bass.

Wohlwill, J. F. (1973). *The study of behavioral development* (pp. 358–375). New York: Academic Press.

IV Analysis of Categorical Longitudinal Data

Models for the Analysis of Change in Discrete Variables*

Chapter 12

CLIFFORD C. CLOGG

Departments of Sociology and Statistics, and
Population Issues Research Center
The Pennsylvania State University
University Park, Pennsylvania

SCOTT R. ELIASON

Department of Sociology
The Pennsylvania State University
University Park, Pennsylvania

JOHN M. GREGO

Department of Statistics
University of South Carolina
Columbia, South Carolina

Abstract

This chapter surveys some log-linear or logit models that can be used to analyze change in discrete endogenous variables observed in a panel-data format. We begin with a new look at the 2×2 contingency table involving repeated measures of a dichotomous variable, including a special design matrix that can be used to study shifts in the marginal distribution as well as association or dependence. Flexible methods for analyzing three generalizations of this simplest case are considered, including (1) models for polytomous variables, (2) models for three or

* Clogg's research was supported in part by Grant No. SES-8709254 from the Division of Social and Economic Sciences of the National Science Foundation. Address correspondence to the first author.

more occasions of measurement, including Markov-type models, and (3) models that allow for covariate effects. The goal is to show how existing methods for the analysis of logit models or more general log-linear models for the cell frequencies can be used to study processes of change. These methods can be applied with computer software that is widely available.

1. Introduction

This chapter deals with models for the analysis of change in discrete variables that are viewed as responses or endogenous variables. The goal is to show that standard techniques for the analysis of discrete data can be used to describe and explain change in such variables. Data are assumed to arise from conventional panel studies with two or more waves. The present development follows standard contingency table models considered in Bishop, Fienberg, and Holland (1975), Haberman (1978, 1979), and Goodman (1984). However, the methods considered here exploit the temporal ordering of repeated measures and are designed for the analysis of growth, development, aging, or similar behavioral concepts. Familiarity with standard log-linear models is assumed; see Fienberg (1980), Knoke and Burke (1980), Clogg and Shockey (1984), Agresti (1984), and Clogg and Eliason (1987/1988) for additional survey material and references to the literature.

We consider the case where T repeated measures of the same discrete response variable Y are available for a sample of N individuals. The observation at time t will be called Y_t. Although we consider cases with time intervals equally spaced, unequal time intervals present no special difficulty with most of the models considered. We suppose that there are $I \geq 2$ levels of Y for each occasion of measurement. Categories of Y may be nominal, ordinal, or quantitative. Although we do not explicitly consider censoring, truncation, or other types of missing data problems, in most cases the models can be modified to incorporate missing observations, regarded as another category of Y at each occasion, in relatively straightforward ways (see Clogg, 1984). We consider only the situation where there is a small or moderate number of repeated measurements (T); when there are many repeated measures, more complicated methods based on quasi-likelihood or other methods should be considered. See Ware and Lipsitz (1988) as well as the other articles in the January 1988 issue of *Statistics in Medicine* for an overview of these other methods.

The organization of this chapter is as follows. First we consider the 2×2 and the general $I \times I$ contingency table that capture all of the information available for a two-wave panel (two occasions) for a dichotomous and a polytomous variable. The aim here is to exploit existing models, as well as reparameterizations of existing models, so that they can be used to answer questions about change in Y. We focus on both the changing distribution of Y over time (marginal distributions) as well as the process by which this change takes place. To characterize the process of change, transition rates and association between categories of Y over time must be considered. Next we consider the natural extension of this problem to three-wave panels (three occasions). A special case of this is the analysis of Markov chains or Markov-like models. Finally, we consider the introduction of explanatory covariates that are *fixed* over time: X variables that are determined at the time of the initial measurement. Both discrete and continuous covariates can be added to these models in straightforward ways. When repeated measures derive from a multiwave panel, having from two to five waves, there is thus a rich array of possible models that can be used to explain change processes.

2. Studying Change with a 2×2 Contingency Table

Let Y_1 and Y_2 denote the measures of dichotomous variable Y at times 1 and 2. When cross-classified with a sample of N individuals, the available information takes the form of a 2×2 contingency table. Let f_{ij} denote the observed frequency in cell (i, j), $i = 1, 2$, $j = 1, 2$, where by convention the first subscript indexes levels of the first measurement (Y_1) and the second subscript indexes levels of the second measurement (Y_2). We let F_{ij} denote the expected frequency under some model, and $\hat{F}_{ij}$ is the sample estimate (or estimator) of F_{ij}. The standard log-linear model (or *log-frequency model*) for this case is

$$\log(F_{ij}) = \lambda + \lambda_{1(i)} + \lambda_{2(j)} + \lambda_{12(ij)}. \tag{1}$$

Numerical subscripts refer to variables Y_1 and Y_2, to remind us of the temporal ordering, and the subscripts in parentheses refer to levels of the variables. When the parameters are defined in the usual way ($\lambda_{1(2)} = -\lambda_{1(1)}$, $\lambda_{2(2)} = -\lambda_{2(1)}$, $\lambda_{12(11)} = \lambda_{12(22)} = -\lambda_{12(12)} = -\lambda_{12(21)}$), we find that $\lambda_{12(11)} = [\log(\theta)]/4$, where θ is the odds ratio, $F_{11}F_{22}/F_{12}F_{21}$. The odds ratio has a long history as a measure of association; see Becker and

Clogg (1988) and references cited there. $\theta = 1$ if and only if Y_1 and Y_2 are independent, in which case the interaction parameters $\lambda_{12(ij)} = 0$ for all i and j. Most literature on contingency tables begins with the odds ratio, or equivalently, with the model in (1). However, association between Y_1 and Y_2 is expected simply because these two quantities are repeated measures of the same variable. Of course, it is important to estimate and analyze this association in terms of θ, $\log(\theta)$, or the interaction values λ_{12}, but these parameters and this model do not say very much about the *change* that has occurred in Y. It is important to realize that the main effect parameters, $\lambda_{1(i)}$ and $\lambda_{2(j)}$, do not measure the marginal distributions of Y_1 and Y_2 in a direct way, so inferences about changing marginals cannot be based directly on the values of these parameters. Ordinarily, when higher-order interactions between or among variables are included (such as λ_{12}), it is not appropriate to interpret the lower-order relatives of these interactions (such as λ_1 and λ_2) as quantities that describe marginal relationships.

An equivalent expression for the model in (1) is as follows. Let Z_{1i} and Z_{2j} denote indicator variables coded $+1$ when i or $j = 1$ and -1 when i or $j = 2$. The quantity $Z_{1i}Z_{2j}$ denotes interaction; the model can be rewritten as

$$\log(F_{ij}) = \lambda + \lambda_{1(1)}Z_{1i} + \lambda_{2(1)}Z_{2j} + \lambda_{12(11)}(Z_{1i}Z_{2j}), \tag{2}$$

which now contains only the four nonredundant parameter values. A common way to exploit the temporal ordering of the variables is to regard Y_2 as a specified response variable and Y_1 as a conditionally fixed explanatory variable, since it occurs prior to Y_2. The ordinary *logit model* (or log-odds model) is equivalent to a reparameterized version of the log-linear model for the frequencies. This model is

$$\Phi_{2(i)} = \log\left(\frac{F_{i1}}{F_{i2}}\right) = 2\lambda_{2(1)} + 2\lambda_{12(i1)} = \beta_2 + \beta_{2\,|\,i}, \tag{3}$$

which can be derived from either (1) or (2). A key quantity in this model is the logit of Y_2; the expression shows how this logit varies with levels of Y_1. Note that the dependence of Y_2 on Y_1 is still measured in terms of $\log(\theta)$, since $\beta_{2\,|\,1} = 2\lambda_{12(11)} = [\log(\theta)]/2$. The logit model is appropriate for analyzing the possible dependence of Y_2 (in terms of the logit) on the levels i of Y_1, but it does not tell us a great deal about change per se.

To study change we must reconsider how the 2×2 table summarizes the available information and then formulate somewhat different models. Cell $(1, 1)$ denotes *persistence* (lack of change) in the sense that f_{11} is the number of persons who started at level 1 of Y and would up at the same level at time 2. Cell $(2, 2)$ likewise represents persistence in state 2. Cell $(1, 2)$ denotes change from state 1 to state 2, and cell $(2, 1)$ denotes change from state 2 to state 1. The magnitudes of F_{12} and F_{21} should be used to describe the change that has occurred, but we will usually also be interested in the marginal distribution of Y_1 (F_{1+}, F_{2+}) and the marginal distribution of Y_2 (F_{+1}, F_{+2}). The fact of the matter is that the parameters of the log-linear model or of the logit model say very little about the comparison of F_{12} and F_{21} or of the two marginal distributions. (The main effect terms, $\lambda_{1(i)}$ and $\lambda_{2(j)}$, do not describe the two marginal distributions except when independence holds $(\lambda_{12(ij)} = 0)$, or in some other quite special circumstances.) For the 2×2 table, however, it is easy to see that change in the marginal distribution over time is directly linked to the magnitudes of F_{12} and F_{21}: we obtain $F_{1+} = F_{+1}$ if and only if $F_{12} = F_{21}$.

To study change more directly, the hypothesis of symmetry $(F_{12} = F_{21})$ and the equivalent hypothesis of marginal homogeneity $(F_{i+} = F_{+i},$ $i = 1, 2)$ are more relevant than the usual log-linear parameterization, which models association or dependence in terms of the odds ratio θ. We note that McNemar's test for symmetry (or marginal homogeneity) is just a test that the binomial parameter $\pi = .5$ in the dichotomy formed from the observed frequencies f_{12} ("successes") and f_{21} ("failures"); see Bishop et al. (1975). However, a model is required to extend the simplest case to more general settings. Two models that direct attention to the analysis of change are as follows. The first is

$$\log(F_{ij}) = \lambda + \lambda_S Z_{ij}^S + \lambda_P Z_{ij}^P + \lambda_A(Z_{1i}Z_{2j}). \tag{4}$$

The superscripts or subscripts S, P, and A refer to *symmetry, persistence,* and *association*, respectively. Z_{1i} and Z_{2j} are defined as with (2). Z_{ij}^S is an indicator variable for the cells (i, j) that represent change; it takes on the values 0, 1, -1, and 0 for cells $(1, 1)$, $(1, 2)$, $(2, 1)$, and $(2, 2)$, respectively. Z_{ij}^P is an indicator variable for cells that represent persistence; it takes on the levels 1, 0, 0, -1, for cells $(1, 1)$, $(1, 2)$, $(2, 1)$, and $(2, 2)$, respectively. Note that the model matrix implied by (3.3) is

$$X^* = \begin{bmatrix} 1 & 0 & 1 & 1 \\ 1 & 1 & 0 & -1 \\ 1 & -1 & 0 & -1 \\ 1 & 0 & -1 & 1 \end{bmatrix} \quad \text{compared with the usual} \quad X = \begin{bmatrix} 1 & 1 & 1 & 1 \\ 1 & -1 & 1 & -1 \\ 1 & 1 & -1 & -1 \\ 1 & -1 & -1 & 1 \end{bmatrix}.$$

Straightforward algebraic manipulation gives $\lambda_S = [\log(F_{12}/F_{21})]/2$, $\lambda_P =$

$[\log(F_{11}/F_{22})]/2$, and $\lambda_A = [\log(F_{11}F_{22}/F_{12}F_{21})]/4 = [\log(\theta)]/4$. Note that $\lambda_S = 0$ if and only if the 2×2 table is symmetric, in which case marginal homogeneity obtains; $\lambda_P = 0$ if and only if the persistence in state 1 is the same as the persistence in state 2; and $\lambda_A = 0$ if and only if Y_1 and Y_2 are independent. The reparameterized model in (3) thus gives three meaningful quantities that can be used to characterize the change in Y. (For later reference, note that covariates can be added to the model easily.) For example, we can consider interactions of any covariate with the indicator variables used to define effects. For closely related results, see Sobel, Hout, and Duncan (1985) and Reiser, Wallace, and Schuessler (1986).

   Let Φ_t denote the log-odds (or logit) for Y at time t, $\Phi_1 = \log(F_{1+}/F_{2+})$, $\Phi_2 = \log(F_{+1}/F_{+2})$. Under marginal homogeneity (or symmetry), we obtain $\Phi_2 = \Phi_1$ (in which case $\lambda_S = 0$ in the previous model); under the condition where the levels of Y are equiprobable, both logits are zero. A model exploiting this information links the two marginal logits together:

$$\Phi_2 = \rho\Phi_1. \tag{5}$$

Note that when $\Phi_1 \neq 0$ (Y_1 marginal is not equiprobable), the parameter $\rho = \Phi_2/\Phi_1$. Note that $\rho = 1$ if and only if marginal homogeneity (or, equivalently, symmetry) holds, ignoring the case where marginals on either or both variables are equiprobable. The maximum likelihood estimator of ρ is $\hat\rho = \log(f_{1+}/f_{2+})/\log(f_{+1}/f_{+2})$, which exists whenever $f_{+1} \neq f_{+2}$. The utility of this formulation of marginal homogeneity can best be seen when covariates are used to predict Φ_1 terms of a linear (logit) model. Suppose $\Phi_1 = \beta_0 + \beta_1 X_1 + \beta_2 X_2$. If the fixed covariates affect the distribution at time 2 only through the distribution at time 1 (i.e., if Y_2 is independent of X_1 and X_2 given Y_1), then $\Phi_2 = \rho(\beta_0 + \beta_1 X_1 + \beta_2 X_2)$, and $\rho\beta_1$ and $\rho\beta_2$ repersent the indirect effects of X_1 and X_2 on Φ_2, respectively. Of course, more general models can be developed. Note that we could have written the model in a slightly different form, $\Phi_2 = \alpha + \Phi_1$, and here α is defined for all possible values of the logits; $\hat\alpha = \log(f_{+1}/f_{+2}) - \log(f_{1+}/f_{2+})$. This parameterization leads to another decomposition of effects, with $\alpha + \beta_1$ and $\alpha + \beta_2$ representing the indirect effects of interest in the situation referred to above.

2.1. Example: Change in Labor Force Status

Table 12.1 gives a 2×2 "mobility table" obtained by cross-classifying variable Y (1 = not employed, 2 = employed) for years $t = 1 = 1984$ and $t = 2 = 1985$. The model of (4) gives $\hat\lambda_S = .025$ ($z = \hat\lambda_S/s(\hat\lambda_S) = .48$), so

TABLE 12.1. Dichotomous Version of Employment
Status, 1984 and 1985 (2×2 table)

1984 ($t = 1$) Status	1985 ($t = 2$) Status[a]		
	Not employed	Employed	Total
Not employed	348	183	531
	(.655)	(.345)	(.255)
Employed	174	1379	1553
	(.112)	(.888)	(.745)
Total	522	1379	2084
	(.250)	(.750)	(1.0)

Source: National Longitudinal Survey of Labor Market Experiences
of Youth. Observations here and elsewhere refer to all persons of ages
21 and 22 in 1979 (the initial wave of the survey).
[a] Row proportions (transition rates) in parentheses for cell frequencies; quantities in parentheses for marginal distributions.

symmetry and marginal homogeneity can be assumed; note that $\hat{\lambda}_S = [\log(f_{12}/f_{21})]/2 = [\log(183/174)]/2$. Also, $\hat{\lambda}_P = -.688$ ($z = -22.95$), which indicates that there are many fewer persons staying in the "not employed" status compared to persons staying in the "employed" status. The association between Y_1 and Y_2 is substantial: $\hat{\lambda}_A = .678$ ($z = 22.29$), or $\log(\hat{\theta}) = 2.713$ and $\hat{\theta} = 15.07$. Transitions to and from employment practically cancel each other out; this is shown clearly by the value of $\hat{\lambda}_S$. The value of $\hat{\rho}$ in model (3.4) is 1.021 (not significantly different from the null value of 1), which is another indication of lack of change in the marginals.

The fitted logit model is $\hat{\Phi}_{2|i} = .714 + 1.356 Z_{1i}$, giving $\hat{\Phi}_{2|i=1} = 2.070$ and $\hat{\Phi}_{2|i=2} = -.642$. This model actually parameterizes the transition rates $r_{ij} = F_{ij}/F_{i+}$; note that the odds $F_{i1}/F_{i2} = r_{i1}/r_{i2}$. The logit model serves well as a model of transition rates, but it is not possible to see from this model that the process is essentially in equilibrium (homogeneous marginals). Notice that this example makes it clear that we should study marginal shifts as well as interaction patterns (differences in transition rates). Model (4) allows for both analyses, whereas model (5) focuses attention on the marginals alone.

3. Studying Change with an $I \times I$ Contingency Table

An $I \times I$ contingency table, for $I > 2$, contains much more information about processes of change or outcomes of change than is the case for a

2×2 table. As before, we let F_{ij} denote the expected frequency in cell (i, j) under some model, with row levels referring to Y_1 and column levels referring to Y_2. F_{i+} and F_{+j} are the marginal distributions at times 1 and 2, respectively. The log-linear model in (1) is still relevant (with subscripts ranging from 1 to I); independence obtains when $\lambda_{12(ij)} = 0$ for all i and j. We rarely expect the independence model to hold when repeated measures are studied, so the problem is again parameterizing the model so that inferences about change are possible. We now give some standard models that can be used for this purpose, including reparameterizations where necessary.

3.1. Marginal Homogeneity, Symmetry, Quasi-Symmetry, and Quasi-Independence

Several special models for square contingency tables serve as the point of departure. *Marginal homogeneity* (MH) says that $F_{i+} = F_{+i}$, $i = 1, \ldots, I$. Because $\sum_i F_{i+} = \sum_i F_{+i} = N$, there are $I - 1$ degrees of freedom available for testing MH. This model is attractive in the repeated measures setting simply because it tests whether there is any change in the distribution of Y over time. However, MH is actually a *linear* rather than a log-linear model: the quantities, $\sum_i F_{ik} - \sum_i F_{ki}$, $k = 1, \ldots, I$, are compared to the null value of 0 in order to test MH (see, e.g., Forthofer and Lehnen, 1981).

To consider the other standard models, we reconsider the log-linear model for the frequencies

$$\log(F_{ij}) = \lambda + \lambda_{1(i)} + \lambda_{2(j)} + \lambda_{12(ij)}. \tag{6}$$

Quasi-symmetry (QS) holds when $\lambda_{12(ij)} = \lambda_{12(ji)}$. To see what this implies, consider the odds ratio

$$\theta_{ij(i'j')} = \frac{F_{ij}F_{i'j'}}{F_{i'j}F_{ij'}}, \tag{7}$$

in the subtable formed from rows i and i' and columns j and j'. It can be verified that $\log(\theta_{ij(i'j')}) = \lambda_{ij} - \lambda_{i'j} - \lambda_{ij'} + \lambda_{i'j'}$. Under QS $\theta_{ij(i'j')} = \theta_{i'j'(ij)}$, or the association is symmetric. It can be verified that the QS model has $(I - 1)(I - 2)/2$ degrees of freedom. *Symmetry* (S) holds when $\lambda_{1(i)} = \lambda_{2(i)}$ and $\lambda_{12(ij)} = \lambda_{12(ji)}$. Under S we obtain $\log(F_{ij}) = \log(F_{ji})$ and $F_{ij} = F_{ji}$. Note that S implies MH. Model S has $I(I - 1)/2$ degrees of freedom. It can also be verified that $S = MH \cap QS$, with $df(S) = I(I - 1)/2 = df(MH) + df(QS) = (I - 1) + (I - 1)(I - 2)/2$. This fact is related to the usual conditional tests of MH in the literature. Let L^2 denote the likelihood-ratio chi-squared statistic for a particular model. Then $L^2(S) - L^2(QS)$ is a conditional test statistic (assuming that QS is true) for MH. The condition that must be satisfied for the validity of the test is the truth of QS. In circumstances where QS is decidedly not true, it is best to use the unconditional test of MH rather than the conditional one.

Although symmetry arises from QS by equating row and column main effects, it is not in general possible to interpret main effects in the QS model as measures of marginal distributions. This should not come as a surprise, because main effect terms do not refer to marginal effects whenever higher-order interactions exist. Consider the situation where $\lambda_{1(i)} = \{-.2, .3, -.1\}$, $\lambda_{2(j)} = \{.4, -.3, -.1\}$, $\lambda_{12(1j)} = \{.3, -.2, -.1\}$, $\lambda_{12(2j)} = \{-.2, .5, -.3\}$, and $\lambda_{12(3j)} = \{-.1, -.3, .4\}$. It is easy to see that $\lambda_{12(ij)} = \lambda_{12(ji)}$, so QS holds. With the constant $\lambda = 1$, we obtain $\log(F_{ij}) = \{1.5, .3, .6; 1.5, 1.5, .9; 1.2, .3, 1.2\}$, and $F_{ij} = \{4.48, 1.35, 1.82; 4.48, 4.48, 2.45; 3.32, 1.35, 3.32\}$. The row marginals are $F_{i+} = \{7.65, 11.41, 7.99\}$ and the column marginals are $F_{+j} = \{12.28, 7.18, 7.59\}$. Note that the departure from marginal homogeneity can in this case be inferred from the main effect parameters, but also that heterogeneity is not a simple function of those parameters.

Quasi-independence (QI) (Goodman, 1968) is another standard model that is useful for analysis of repeated measures. In the present context, diagonal entries (indicating persistence in states over time) might be singled out for special treatment. Model (7) can be replaced by

$$\log(F_{ij}) = \lambda + \lambda_{1(i)} + \lambda_{2(j)} + \lambda_{12(i)}, \tag{8}$$

where now a single parameter $\lambda_{12(i)}$ is used for each diagonal cell, $i = 1, \ldots, I$. Written in this form, it is easy to see that QI is a special model for interaction. In fact, for cells not on the main diagonal, we obtain $\log(F_{ij}) = \lambda + \lambda_{1(i)} + \lambda_{2(j)}$, or Y_1 and Y_2 are conditionally independent off the diagonal. In other words, movement is random for those who move under the QI model. It can be verified that QI implies QS, so a partitioning strategy leading to conditional tests can be used.

Several interesting reparameterizations of the foregoing models applied to mobility tables can be found in Sobel, Hout, and Duncan (1985) and Hout, Duncan, and Sobel (1987). We note, for example, that the reparameterization of the model for the 2×2 table used in Section 2 can be generalized for this setting.

We next consider a useful reparameterization of marginal homogeneity that is well suited for the analysis of marginal shifts in Y. Let

$\Phi_{1(i)} = \log(F_{i+}/F_{I+})$ and $\Phi_{2(i)} = \log(F_{+i}/F_{+I})$, for $i = 1, \ldots, I - 1$, denote the $I - 1$ *marginal* logits for the Y_1 and Y_2 variables. In place of the model in (5), consider the model

$$\Phi_{2(i)} = \rho_i \Phi_{1(i)}. \tag{9}$$

Note that MH obtains (unconditionally) if and only if $\rho_i = 1$ for $i = 1, \ldots, I - 1$, ignoring some boundary constraints (where $\Phi_{1(i)} = 0$). The utility of this model is simply that it links the two marginal distributions together. If Y_1 were to be linked to a set of covariates via a multinomial logit model, $\Phi_{1(i)} = \beta_i + \beta_{1i}X_1 + \beta_i X_2$, then $\rho_i \beta_{1i}$ and $\rho_i \beta_{2i}$ are the indirect effects of interest. If we are interested in analyzing change in Y over time, this model might be more informative than any of the models already discussed. In fact, it might be argued that this new model is the only one considered thus far that actually deals with change in Y per se.

3.2. An Example: Change in Labor Force Status (Four States)

Table 12.2 gives a 4×4 table analogous to Table 12.1, where Y now has four categories (1 = not in labor force, 2 = unemployed, 3 = part-time or absent from work, 4 = full-time employed). Again, Y_1 is the 1984

TABLE 12.2.   Employment Status in 1984 and 1985 (4×4 table)

1984 ($t = 1$) Status	Not in labor force	Unemployed	Part-time	Full-time	Total
	1985 ($t = 2$) Status				
Not in LF	215	45	61	35	356
	(.604)	(.126)	(.171)	(.098)	(.171)
Unemployed	36	52	37	50	175
	(.206)	(.297)	(.211)	(.286)	(.084)
Part-time	60	29	160	194	443
	(.135)	(.065)	(.361)	(.438)	(.214)
Full-time	43	42	189	836	1110
	(.039)	(.038)	(.170)	(.753)	(.533)
Total	354	168	447	1115	2084
	(.170)	(.081)	(.214)	(.535)	(1.0)

Source: Same as Table 12.1. row proportions (transition rates) in parentheses; row marginal distribution expressed as proportion of total.

TABLE 12.3. Chi-squared Statistics for Some Models Applied to the Data in Table 12.2

Model	Degrees of freedom	Chi-squared[a] statistic
Independence	9	869.19
Marginal homogeneity	3	.28
Symmetry	6	3.57
Quasi-symmetry	3	3.29
Quasi-independence	5	81.55

[a] For MH model, a Wald statistic is used, whereas the other values are likelihood-ratio statistics.

measure, and Y_2 is the 1985 measure. The categories might actually be viewed as ordered, but at present we do not exploit this fact. It is readily apparent that the marginals are remarkably stable over time, in spite of substantial shifts in labor force status over the course of a year. Well over half (1115/2084) of the individuals were full-time employed in both waves.

Table 12.3 gives the fit statistics for models applied to the data in Table 12.2. Y_1 and Y_2 are not independent ($L^2 = 886.19$ for the independence model), but the marginals are clearly quite homogeneous (W = Wald statistic = .28 on 3 df). Symmetry and quasi-symmetry are also viable models for the data; the former indicates that shifts were indeed compensating shifts (f_{ij} approximately equal to f_{ji} for all $i \neq j$), and both indicate that the association is symmetric. The QI model with a special parameter for each diagonal cell does not fit ($L^2 = 81.55$ on 5 df); but the reduction in L^2 given by $L^2(I) - L^2(QI) = 787.64$ (4 df) shows that about 91% of the row-column association is due to state persistence. A relatively small fraction of the overall association can be traced to off-diagonal cells that represent change in labor force status.

The conditional test of MH is $L^2(S) - L^2(QS) = 3.57 - 3.29 = .28$ (3 df), which in this case is the same (to two decimal places) as the chi-squared value for the unconditional test of MH. (It must be emphasized that the equivalence of the two test statistics is simply an accident. The two procedures need not agree so closely in circumstances where QS does not hold.)

The ρ_i parameters in the model of (9) are estimated as .991, .976, 1.005, for $i = 1, 2, 3$. The test for MH indicates that these values are not significantly different from 1. The (nearly) symmetric off-diagonal entries (including nearly symmetric association) produces approximate equality in the marginals, but the ρ parameters lead to quantities that are easy to incorporate in more general models, as discussed next.

4. Studying Change with a $2 \times 2 \times 2$ Table

Consideration of the cross-classification of Y_1, Y_2, and Y_3 for the case of dichotomous measures suggests some generalizations appropriate for T repeated measures (T moderate, say less than 5). We start with the standard log-linear model for the 2^3 table,

$$\log(F_{ijk}) = \lambda + \lambda_{1(i)} + \lambda_{2(j)} + \lambda_{3(k)} + \lambda_{12(ij)}$$
$$+ \lambda_{13(ik)} + \lambda_{23(jk)} + \lambda_{123(ijk)}. \tag{10}$$

It is well known that

$$\lambda_{12(11)} = \tfrac{1}{8}\left[\log\left(\frac{F_{111}F_{221}}{F_{121}F_{211}}\right) + \log\left(\frac{F_{112}F_{222}}{F_{122}F_{212}}\right)\right]$$
$$= \frac{1}{4}\frac{\log(\theta_{12(1)}) + \log(\theta_{12(2)})}{2},$$

where $\theta_{12(k)}$ is the odds-ratio (or conditional odds-ratio) between Y_1 and Y_2 when Y_3 is at level k. The quantity in braces is usually called the *partial* log-odds-ratio, which we represent as $\log(\theta_{12.3})$. (This refers to the odds-ratio between Y_1 and Y_2 adjusted for Y_3.) Similar expressions follow for the λ_{13} and λ_{23} parameters, which are functions of $\theta_{13.2}$ and $\theta_{23.1}$, respectively. Note that typically we will only be interested in λ_{13} and λ_{23} (or $\theta_{13.2}$ and $\theta_{23.1}$) and not in λ_{12} (or $\theta_{12.3}$), simply because the latter gives the association between the two *antecedent* variables, Y_1 and Y_2, adjusting for the *consequent* variable Y_3. Several special cases or reparameterizations of this model are useful for studying change.

4.1. Standard Hierarchical Models

The usual log-linear models for three-way tables arise by considering the obvious restricted versions of the saturated model given in (10). The independence model arises by setting all two-factor interactions plus the three-factor interaction at zero. The model of no three-factor interaction arises by setting just the three-factor interaction to zero. For this model, it is natural to consider the logit of Y_3 expressed in terms of the levels of Y_1 and Y_2, that is,

$$\Phi_{3(ij)} = \log\left(\frac{F_{ij1}}{F_{ij2}}\right) = \beta_3 + \beta_{3\,|\,i} + \beta_{3\,|\,j}, \tag{11}$$

where $\beta_3 = 2\lambda_{3(1)}$, $\beta_{3\,|\,i} = 2\lambda_{13(i1)}$, and $\beta_{3\,|\,j} = 2\lambda_{23(j1)}$. It follows that $\beta_{3\,|\,i=1} = \{\log(\theta_{13.2})\}/2$ and $\beta_{3\,|\,j=1} = \{\log(\theta_{23.1})\}/2$, so the two partial log-odds-ratios become the essential parameters in this additive logit model. Note that $\beta_{3\,|\,i}$ measures the "lagged" effect of Y_1 on the log-odds for Y_3 (adjusted for the intervening variable, Y_2), and that $\beta_{3\,|\,j}$ measures the effect of Y_2 on the log-odds of Y_3 (adjusted for the antecedent variable, Y_1). The additive logit model in (11), which is equivalent to the model of no three-factor interaction, is certainly a reasonable model for summarizing association between Y_3 and $Y_1 - Y_2$, but as before we need to consider reparameterizations in order to focus on the process of change per se.

4.2. A Different Approach Based on a Special Design Matrix

The log-frequency model in (10) can be written in terms of indicator variables as follows:

$$\begin{aligned}\log(F_{ijk}) = \lambda &+ \lambda_{1(1)}Z_{1i} + \lambda_{2(1)}Z_{2j} + \lambda_{3(1)}Z_{3k} + \lambda_{12(11)}(Z_{1i}Z_{2j}) \\ &+ \lambda_{13(11)}(Z_{1i}Z_{3k}) + \lambda_{23(11)}(Z_{2j}Z_{3k}) + \lambda_{123(111)}(Z_{1i}Z_{2j}Z_{3k}), \end{aligned} \tag{12}$$

where $Z_{1i} = +1$ if $i = 1$, $Z_{1i} = -1$ if $i = 2$, with similar definitions for Z_{2j} and Z_{3k}. Note that only nonredundant parameter values are included explicitly; there are eight such parameters, counting the constant term. Because of the temporal ordering of Y_1, Y_2, and Y_3, there are, however, only *three* parameters of direct interest: λ_{13}, λ_{23}, and λ_{123}, the interactions involving Y_3. This is the case because (a) we are not interested in the main effects when interactions are present (in the usual design anyway) and (b) we are not interested in the association between the temporally prior variables Y_1 and Y_2 adjusted for levels of Y_3 (captured in the λ_{12} parameter). A reparameterization that maintains the association parameters of interest but incorporates other parameters useful for studying change is as follows.

Let $\nu_{ijk} = \log(F_{ijk})$ and $\mathbf{v} = \{\nu_{ijk}\}$, a vector. Any log-linear model can be represented as $\mathbf{v} = Z\boldsymbol{\lambda}$, where Z is the design matrix and $\boldsymbol{\lambda}$ is the vector of parameters. We consider the special design matrix Z displayed in Table 12.4. The columns of this design matrix correspond to the parameters in

TABLE 12.4. A Special Design Matrix for the $2 \times 2 \times 2$ Table

Cell (i, j, k)	Design matrix							
	1	2	3	4	5	6	7	8
(1, 1, 1)	1	1	0	1	1	0	0	0
(1, 1, 2)	1	1	1	0	−1	0	0	0
(1, 2, 1)	1	1	−1	0	−1	0	0	0
(1, 2, 2)	1	1	0	−1	1	0	0	0
(2, 1, 1)	1	−1	0	0	0	0	1	1
(2, 1, 2)	1	−1	0	0	0	1	0	−1
(2, 2, 1)	1	−1	0	0	0	−1	0	−1
(2, 2, 2)	1	−1	0	0	0	0	−1	1

the model; column 1 corresponds to the constant term λ. Column 2 corresponds to the main effect of Y_1; the column vector here is just Z_{1i} as before. Columns 3–8 replace the vectors in the usual design matrix and they have the following interpretations:

Column 3: A contrast between the $(1, 1, 2)$ and the $(1, 2, 1)$ cells, giving a parameter $\lambda_{S(1)}$ measuring asymmetry between Y_2 and Y_3 when Y_1 is at level 1.

Column 4: A contrast between the $(1, 1, 1)$ and the $(1, 2, 2)$ cells, giving a parameter $\lambda_{P(1)}$ measuring the persistence between Y_2 and Y_3 when Y_1 is at level 1;

Column 5: A contrast that measures association between Y_2 and Y_3 when Y_1 is at level 1, with parameter $\lambda_{A23(1)}$ defined so that $4\lambda_{A23(1)} = \log(\theta_{23(1)})$.

Columns 6–8: Contrasts corresponding to those in columns 3–5, respectively, measuring asymmetry, persistence, and association when Y_1 is at level 2, with parameters $\lambda_{S(2)}$, $\lambda_{P(2)}$, and $\lambda_{A23(2)}$, respectively.

There are many other reparameterizations of the design for the $2 \times 2 \times 2$ table besides this one. But note how the elementary model for the 2×2 table given in (4) was used to motivate this model, and we now have six parameters that can be used to study the quantities of interest. The usual design matrix gives only three parameters of interest, and only those that measure association rather than change. An interesting property of the present design is that estimates of asymmetry effects are uncorrelated with estimates of persistence effects and that estimates of effects for $Y_1 = 1$ are uncorrelated with estimates of effects for $Y_2 = 2$. This property suggests ways to analyze the possible dependence of these general effects on fixed covariates.

4.3. Markov Models

Let $P^{(t)}_{j|i} = P(Y_t = j \mid Y_{t-1} = i)$ denote the *transition rate* or conditional probability that state j is held at time t given that state i is held at time $t - 1$. A first-order Markov chain means that the serial *state dependence* is of the first order, or that $P(Y_t = j \mid Y_{t-1} = i, Y_{t-2} = i', \ldots, Y_1 = i^*) = P(Y_t = j \mid Y_{t-1} = i)$ for all i and j and for all $i', \ldots, i^*$. For $T = 3$, this model implies that Y_1 and Y_3 are conditionally independent given the levels of Y_2, or that the first-order Markov model fits the marginals $\{(Y_1 Y_2), (Y_2 Y_3)\}$. In terms of the log-frequency model of (10), this model is satisfied when $\lambda_{12} = \lambda_{123} = 0$. In terms of the additive logit model in (11), we obtain the first-order Markov model by setting $\beta_{3|i} = 0$. Chi-squared statistics for either version of the model are identical and can be used to test the assumption of first-order dependence. For an $I \times I \times I$ table, this model has $(I - 1)(I - 1)[1 + (I - 1)] = I(I - 1)(I - 1)$ degrees of freedom.

A first-order Markov process is *stationary* when $P^{(t)}_{j|i} = P_{j|i}$, that is, when the transition rates are independent of time. As noted in Bishop, Fienberg, and Holland (1975), pp. 265–267), stationarity (in the context of a first-order process) can be tested by forming an $I \times I \times (T - 1)$ table from the adjacent pairs of transition matrices or contingency tables, or a $2 \times 2 \times 2$ table when studying stationarity in the context of a $2 \times 2 \times 2$ table of repeated measures. The first $I \times I$ table is $Y_1 \times Y_2$, the second is $Y_2 \times Y_3$, etc., for $T - 1$ such tables. Stationarity in the above sense holds if and only if the model $\{(AB), (AC)\}$ holds, where A refers to the first variable (initial state), B refers to the second variable (ending state), and C refers to time. For this model applied to the $I \times I \times I$ table, there are $(I - 1)(T - 2) + (I - 1)(I - 1)(T - 2) = I(I - 1)(T - 2)$ degrees of freedom.

We test the Markov chain assumption by considering the two separate components given above, one for testing the order and one for testing stationarity. Summing the two test statistics gives an overall test statistic, but it will often be useful to consider the magnitude of each component separately to examine possible departures from assumptions. Note that when there are more than two states ($I > 2$), there are many possible ways to modify the basic Markov model using standard log-linear models, for either or both component models. We indicate later that it is also possible to introduce covariates in Markov chains using similar logic.

4.4. Marginal Homogeneity

When a Markov process is in equilibrium, marginal distributions are constant over time. Note that a stationary process leads to constant marginals with sufficiently large T, but for any given empirical setting the marginals can be quite heterogeneous unless the process has been operating for some time prior to the periods of observation. It is natural therefore to consider the generalization of marginal homogeneity for the $2 \times 2 \times 2$ or $I \times I \times I$ table, or for even more repeated measures. As before, this model is a *linear* hypothesis given by

$$\sum_{i,j} F_{ijk} = \sum_{j,k} F_{jik} = \sum_{j,k} F_{jki}, \qquad i = 1, \ldots, I. \tag{13}$$

Because each marginal sums to N, there are at most $3(I-1)$ unique marginal counts $[T(I-1)$ for the I^T table], so there are $3(I-1) - (I-1) = 2(I-1)$ degrees of freedom for MH [and $(T-1)(I-1)$ degrees of freedom for the I^T table of repeated measures]. The MH hypothesis can be of interest in the context of Markov chains in order to detect how close the process is to the steady state or equilibrium distribution, and generalizations of MH are useful, as in Section 2, to characterize the change in marginal distributions over time.

4.5. Example: Change in Labor Force Status Over Three Waves

Table 12.5 presents a $2 \times 2 \times 2$ table with dichotomous Y values, for $t = 1 = 1983$, $t = 2 = 1984$, and $t = 3 = 1985$, $Y_t = 1$ for not employed and $Y_t = 2$ for employed. The estimated transition rates for 1983 to 1984 and for 1984 to 1985 also appear in this table; it is clear that the transition rates are quite homogeneous over time, but of course it is not possible to see from these estimates whether the order of the process is first-order. Table 12.6 gives chi-squared values and degrees of freedom for several models discussed above applied to this table.

There is obviously a great deal of dependence among the three variables, as indicated by the L^2 value for independence of 1324.98 on 4 df. The first-order Markov model does not fit ($L^2 = 102.50$ on 2 df), whereas the process is reasonably stationary ($L^2 = 1.09$ on 2 df). The test

TABLE 12.5. Dichotomous Version of Employment Status, 1983, 1984, 1985 ($2 \times 2 \times 2$ table)

1984 ($t = 2$) Status	The $2 \times 2 \times 2$ contingency table 1985 ($t = 3$) Status[a]		
	Not employed	Employed	Total
1983 ($t = 1$) status: not employed			
Not employed	281	96	
Employed	56	132	
1983 ($t = 1$) status: employed			
Not employed	67	87	
Employed	118	1247	

Estimated transition rates ($\hat{P}^{(t)}_{j|i}$)

Transitions from $t = 1$ to $t = 2$
.667	.333
.101	.899

Transitions from $t = 2$ to $t = 3$
.655	.345
.112	.888

TABLE 12.6. Chi-Squared Values for Some Models Applied to the Data in Table 12.5

Model/description	Degrees of freedom	Chi-squared statistic
Independence/ $\{(Y_1), (Y_2), (Y_3)\}$	4	1324.98
First-order Markov/ $\{(Y_1 Y_2), (Y_2 Y_3)\}$	2	102.50
Stationary transition rates	2	1.09
Marginal homogeneity	2	5.21[a]
No three-factor interaction/ $\{(Y_1 Y_2), (Y_2 Y_3), (Y_1, Y_3)\}$	1	.36

[a] Wald statistic used for MH model; likelihood-ratio statistic for all others.

of a first-order, stationary Markov chain is the sum of these two values, $L^2 = 102.50 + 1.09 = 103.59$ on 4 df, which clearly leads to rejection of the model. But the decomposition tells us that the first-order assumption and not the stationarity assumption is at fault. Marginal homogeneity provides a satisfactory fit (Wald statistic $= 5.21$ on 2 df), so whatever the process that generated the data we can say that marginal distributions are essentially at an equilibrium level. Finally, the model that incorporates dependence between Y_1 and Y_3 in addition to the first-order model, which is equivalent to the model of no three-factor interaction, gives an acceptable fit, $L^2 = .36$ on 1 df. This model can be represented in logit form, with fitted equation

$$\hat{\Phi}_{3(ij)} = \log\left(\frac{\hat{F}_{ij1}}{\hat{F}_{ij2}}\right) = -.62 + .71(Z_{1i}) + 1.01(Z_{2j}).$$

The effect of Y_1 on the log-odds of Y_3, adjusted for Y_2, is .71; the corresponding adjusted log-odds-ratio is $2(.71) = 1.42$ (adjusted odds-ratio is $\exp(1.42) = 4.14$), so the lagged effect is substantial. (Note that this quantity would be approximately zero if a first-order Markov model were true.)

Table 12.7 gives the parameter values for the model discussed in Section 4.2 using the special design matrix used to code symmetry, persistence, and association effects. (The parameter corresponding to the constant term is not reported.) Note that we essentially condition on the

TABLE 12.7. Parameter Values for the Model Developed in Section 4.2 (design matrix in Table 12.4), Applied to the Data in Table 12.5

Parameter[a]	Estimate $(= \hat{\lambda})$	$\hat{\lambda}/s(\hat{\lambda})$	Interpretation
2	−.18	−5.32	Main effect of levels i of Y_1
3	.27	3.21	Asymmetry in Y_2 and Y_3 when $Y_1 = 1$
4	.38	7.16	Persistence in Y_2 and Y_3 when $Y_1 = 1$
5	.48	9.73	Association between Y_2 and Y_3 when $Y_1 = 1$
6	−.15	−2.16	Asymmetry in Y_2 and Y_3 when $Y_1 = 2$
7	−1.46	−23.31	Persistence in Y_2 and Y_3 when $Y_2 = 2$
8	.52	11.10	Association between Y_2 and Y_3 when $Y_1 = 2$

[a] Parameters correspond to columns of the design matrix; see Section 4.2.

state at time 1 (Y_1) and examine the relevant effects for Y_2 and Y_3 at each level of Y_1. The model is one natural way to extend the setup given earlier for the 2×2 table using the special design matrix presented for that case. All effects are statistically significant. The asymmetry effect is .27 when $Y_1 = 1$; multiplying by 2 gives the logit of interest. In other words, given that $Y_1 = 1$, there are $\exp(.54) = 1.72$ times more people moving from state 1 to state 2 than in the reverse direction. The asymmetry effect is $-.15$ when $Y_2 = 2$; the two asymmetry effects are significantly different. Persistence effects defined conditionally on Y_1 are both very strong. The association effects are both estimated at about .50 and are not significantly different from each other (no three-factor interaction). The pooled association effect multiplied by four gives the log-odds-ratio, or about 2.00, which corresponds to the first-order lagged effect in a logit model. In summary, the reparameterization used to produce the estimates in Table 12.7 provides a convenient way to summarize change in terms of at least six parameters, whereas earlier approaches essentially estimate lagged associations or model time dependence.

5.   Studying Change with an $I \times I \times I$ Table

Each model considered in Section 4 can be generalized in straightforward ways for the $I \times I \times I$ table. In addition, many special constraints can be imposed on these models to parameterize change processes in a succinct way. The coding of the design matrix given in Section 4.2 can be modified in a variety of ways; see Sobel (1988) for guidelines. Rather than go over specific models for the $I \times I \times I$ table, we proceed immediately to empirical illustrations. Note that adding categories (or disaggregating categories used in the $2 \times 2 \times 2$ case) leads to a method of partitioning variability in almost all cases. We would ordinarily be interested in the comparison of results from $2 \times 2 \times 2$ tables with results from $I \times I \times I$ tables to examine whether the latter provide possible explanations for results obtained from the former. For example, a Markov chain model might be unsatisfactory for the $2 \times 2 \times 2$ table, in terms of either the first-order dependence assumption or the time-stationarity assumption, because of aggregation or "lumping" of states. Such issues can be examined directly by disaggregating states.

5.1. Example: Change in a Four-Fold Labor Force Status Variable Over Three Waves

Table 12.8 presents a $4 \times 4 \times 4$ table with Y categorized into four levels, $1 =$ not in the labor force, $2 =$ unemployed, $3 =$ part-time, $4 =$ full-time. Thus, there are $T = 3$ repeated measures (1983, 1984, and 1985). Note that categories 1–2 of Y here pertain to category 1 of Y (not employed) in Section 4.5, while categories 3–4 of Y here pertain to category 2 of Y in Section 4.5. In other words, both states recognized earlier have been disaggregated into two categories for each. While there are some cases

TABLE 12.8.   Employment Status in 1983, 1984, and 1985 ($4 \times 4 \times 4$ table)

1984 ($t = 2$) Status	1985 ($t = 3$) Status			
	Not in labor force (1)	Unemployed (2)	Part-time (3)	Full-time (4)
1983 ($t = 1$) status = 1 (not in labor force)				
1. Not in LF	162*	18	32	17
2. Unemployed	14	15	6	9
3. Part-time	22	4	7	11
4. Full-time	10	4	13	21
1983 ($t = 1$) status = 2 (unemployed)				
1. Not in LF	23	16	7	5
2. Unemployed	9	24*	12	8
3. Part-time	5	5	12	22
4. Full-time	2	4	9	37
1983 ($t = 1$) status = 3 (part-time)				
1. Not in LF	20	5	13	6
2. Unemployed	10	8	9	15
3. Part-time	20	15	86*	70
4. Full-time	11	9	45	161
1983 ($t = 1$) status = 4 (full-time)				
1. Not in LF	10	6	9	7
2. Unemployed	3	5	10	18
3. Part-time	13	5	55	91
4. Full-time	20	25	122	617*

Note: Asterisk (*) denotes persistence in state over all three waves.

TABLE 12.9.   Chi-Squared Values for Some Models Applied to
the Data in Table 12.8

Model/description	Degrees of freedom	Chi-squared statistic
Independence/		
$\{(Y_1), (Y_2), (Y_3)\}$	54	1980.44
First-order Markov/		
$\{(Y_1Y_2), (Y_2Y_3)\}$	36	70.69
Stationary transition rates	12	14.34
Marginal homogeneity	6	21.00[a]
No three-factor interaction/		
$\{(Y_1Y_2), (Y_2Y_3), (Y_1Y_3)\}$	27	31.12
Markov model with persistence		
effects	32	106.16[b]
Markov model with added		
persistence effects	28	40.61[c]

[a] Wald statistic reported for MH model; likelihood-ratio statistic reported for other models.

[b] First-order Markov model with persistence effects in $(1, 1, 1)$, $(2, 2, 2)$, $(3, 3, 3)$, and $(4, 4, 4)$ cells.

[c] First-order Markov model with persistence effects as in note (b) plus persistence effects for $(1, j, 1)$, $(2, j, 2)$, $(3, j, 3)$, $(4, j, 4)$. The latter effects can be viewed as common parameters for each j (level of Y_2).

where the truth of a model for the disaggregated states implies the truth of the corresponding model for the aggregated states, in general we should think of the two tables (i.e., a $2 \times 2 \times 2$ and a $4 \times 4 \times 4$ table) as leading to substantively different models of the process of change. Table 12.9 gives chi-squared values and degrees of freedom for several models applied to the data in Table 12.8.

The independence model does not describe the data ($L^2 = 1980.44$ on 54 df), which is hardly surprising. As before, the first-order Markov model does not fit the data satisfactorily ($L^2 = 170.69$ on 36 df), but positing only first-order dependence accounts for $(1980.44 - 170.69)/1980.44 = 91.4\%$ of the serial dependence among states. Transition rates are, however, quite stationary ($L^2 = 14.34$ on 12 df), so the Markov chain model fails solely by virtue of the fact that the serial dependence is not first-order. For the three waves considered simultaneously, we have evidence that marginal homogeneity cannot be assumed (chi-squared value for MH is 21.00 on 6 df). Referring to the result in Table 12.3 for MH applied to just the 1984 and 1985 waves, where MH was quite satisfactory, shows that MH fails here mainly because the 1983 ($t = 1$) marginal distribution is quite different from the other two. Given that MH for the 2^3 table was satisfactory (Table 12.6), we see that disaggregation has led to marginal inhomogeneity, principally between the first-wave distribution and the other two waves. The observed marginal distributions are as follows:

$$Y_1 = \{.175, .096, .241, .488\},$$

$$Y_2 = \{.171, .083, .213, .533\},$$

$$Y_3 = \{.170, .081, .214, .535\}.$$

Obviously, the marginals differ between the first and the second two waves primarily in higher levels of unemployment and part-time employment and lower levels of full-time employment. When the logit formulation relating marginal distributions is used (see Section 2), using $Y = 4$ (full-time) as the reference category, we find $\hat{\rho}_i = \{2.018, .880, .765\}$ for the 1983/1984 comparison and $\hat{\rho}_i = \{.991, .976, 1.005\}$ for the 1984/1985 comparison, for Y states $i = 1$, 2, and 3.

The model of no three-factor interaction provides a satisfactory fit ($L^2 = 31.12$ on 27 df), so the serial dependence can be characterized rather simply. In fact, this model can be case in a logit-model form. Let $\Phi_{3(k)\,|\,ij}$ denote the log-odds, $\log(F_{ijk}/F_{ij4})$. We obtain the following system of logit equations directly from the model of no three-factor interaction:

$$\hat{\Phi}_{3(1)\,|\,ij} = -.637 + \hat{\beta}_{(1)i} + \hat{\beta}_{(1)j}$$

$$\hat{\Phi}_{3(2)\,|\,ij} = -1.017 + \hat{\beta}_{(2)i} + \hat{\beta}_{(2)j}$$

$$\hat{\Phi}_{3(3)\,|\,ij} = -.310 + \hat{\beta}_{(3)i} + \hat{\beta}_{(3)j}$$

The $\beta_{(k)i}$ values describe the dependence of the logits of Y_3 on the levels of Y_1 adjusted for levels of Y_2. These quantities are $\hat{\beta}_{(1)i} = \{1.310, -.047, -.223, -1.040\}$, $\hat{\beta}_{(2)i} = \{.472, .635, -.269, -.839\}$, and $\hat{\beta}_{(3)i} = \{.279, -.239, .165, -.322\}$. (Note that these parameter sets need not sum to zero because they refer to contrasts of each k-level with $k = 4$.) The lagged effects are substantial, especially for the first two sets of contrasts.

We next consider two modifications of the first-order Markov model that suggest generalizations for characterizing departures from the assumptions of the model. The first incorporates "diagonal" effects for the full three-way transition matrix. A single parameter is added for each cell indicating three-wave persistence. That is, the three-factor interaction parameter $\lambda_{123(i)}$ is used, for a total of four additional parameters, to account for persistence across the three waves. This model (see Table 12.9) gives $L^2 = 106.16$ on $36 - 4 = 32$ df, so adding these four persistence parameters reduces L^2 by $170.69 - 106.16 = 64.53$ sacrificing only 4 df. Next we add parameters that include persistence in state i over the first and the last waves by adding dummy variables $Z_{ik} = 1$ if $i = k$, 0 otherwise, for a total of four additional parameters. This model gives $L^2 = 40.61$ on $32 - 4 = 28$ df, which fits the data satisfactorily if a conventional .05 level of significance is used (p-value = .058). Other special contrasts or dummy variables can be used to explain the lagged dependence beyond the first-order, but the two kinds of persistence effects added to the first-order model certainly indicate some of the possibilities.

6. Incorporating Fixed Covariates in Models of Change

In this section we consider how time-invariant, or fixed, covariates can be incorporated in models of change. To simplify discussion, we consider only the case where the x-variables do not change over time, or are constant from wave to wave, so that these variables are actually fixed at the time of the initial wave of measurement. Throughout this section we assume that a vector of covariates, $\mathbf{x}^T = \{x_1, x_2, \ldots, x_K\}$, is available. The x-values can consist of discrete variables, which can be coded in a variety of ways (e.g., through the use of dummy variables), continuous variables, or both, and they might consist of interactions as well. Although it is possible to modify these methods to incorporate time-varying covariates, in many cases it is difficult to distinguish between a time-varying covariate and the time-varying response variable (Y) in terms of "causal" or structural representation. In other words, it is often difficult to know whether changes in x-variables are to be structurally related to changes in Y, whether changes in Y are actually structurally related to changes in x-variables, or whether changes in both variables are endogeneous or simultaneous.

6.1. Time-Sequence Logit Models

Goodman's (1973) path-analytic method lends itself easily to studying association between responses over time. Similarities to Markov models will be drawn out below. We now let h, $h = 1, \ldots, N$, denote the subject or case, $Y_{t(h)}$ the response of the hth individual at time t, for $t = 1, \ldots, T$. We first consider the case with dichotomous Y.

A fairly general model that includes covariate effects as well as lagged effects of the response is

$$\Phi_{t(h)} = \log\left[\frac{P(Y_{t(h)} = 1)}{P(Y_{t(h)} = 0)}\right]$$

$$= \alpha + \beta_{t-1}Z_{t-1}, (h) + \cdots + \beta_1 Z_{1(h)} + \gamma_t'\mathbf{x}_{(h)}, \tag{14}$$

where $Z_{t(h)}$ is a dummy variable (or a 1, -1 variable) for $Y_{t(h)}$, $\mathbf{x}_{(h)}$ is the vector of covariate values for the hth subject, and γ is the coefficient vector for the covariate effects. The model can be generalized by including interactions among the lagged response variables (products of the Z's) and/or interactions of the response variables with covariates. The β-parameters measure lagged effects of the response levels on the logit of the response at time t. For T waves there are T sequenced equations and the model consists of the set of T equations; the response at time 1 ($\Phi_{1(h)}$) would include only the covariate effects. Further specifications are possible, such as setting $\gamma_t = \gamma$ (constant covariate effects over the sequence). Usually in considering such models we would wish to examine the *order* of the process (examine the β_{t^*}, for $t^* = 1, \ldots, t - 1$) as well as the degree to which covariate effects persist over time once prior states are controlled. Because this model is merely a set of separate logit regressions, there is no special difficulty in estimating or testing the model. Note that by virtue of the fact that the logit model is a special form of the hierarchical log-linear model, this model generalizes the logit model of Section 2.1 or Section 4.1.

With polytomous responses, the model of (14) can be generalized in several ways. One method is to form contrasts, $\Phi_{t,i(h)} = \log[P(Y_{t(h)} = i)/P(Y_{t(h)} = I)]$, using level I of Y_t as the reference category, and then consider the multinomial-logit model

$$\Phi_{t,i(h)} = \alpha_1 + \beta_{t-1,i,j}Z_{t-1,j(h)} + \cdots + \beta_{1,i,j}Z_{1,j(h)} + \gamma_{t,i}'\mathbf{x}_{(h)}. \tag{15}$$

Here, $Z_{t,j(h)}$ is a dummy variable taking on the value 1 if $Y_{t(h)} = j$, 0 otherwise; importantly, the covariate effects depend on the response contrast employed.

When the polytomous response variable is ordered, there are a variety of possible ways to use time-sequence logit models or logit-type models. See Goodman (1984) and McCullagh (1980) for examples. One approach is to let $\Phi_{t,i(h)} = \log[P(Y_{t(h)} > i)/P(Y_{t(h)} \leq i)]$, for $i = 1, \ldots, I - 1$, which exploits the $I-1$ high-versus-low thresholds that can be formed from I ordered response categories. The lagged effects can be included as a set of dummy variables, as in (15), or even included as linear effects. A model incorporating the latter is

$$\Phi_{t,i(h)} = \alpha_i + \beta_{t-1} Y_{t-1,(h)} + \cdots + \beta_1 Y_{1(h)} + \gamma'_t \mathbf{x}_{(h)}. \tag{16}$$

Here note that there is only one effect for each lagged response (because of the assumption that lagged effects are linear and because only one effect is used for the set of contrasts of the response). This model appears quite satisfactory for analysis of ordinal responses in a time sequence. The usefulness of these three basic models is illustrated next.

6.2. Examples of Time-Sequence Logit Models: Labor Force Status

Tables 12.10 and 12.11 give model estimates for the time-sequence logit models represented by (14) and (16), respectively. [Model (15) is slightly more complicated, because there are so many effects involved. A modified version of this model will be considered later.] We first consider the dichotomous version of Y used previously and three waves (1983, 1984, 1985). The set of three logit equations that comprise the time-sequence logit model include several covariates: SEX, NONBLACK, RU83 (urban/rural residence in 1983), and SOUTH (residence in the South in 1983), all dummy variables, and UNEMR83, a contextual variable (quantitative) giving the unemployment rate in the "local labor market" in which the individual resided in 1983. The logit equation for Y_1 shows that SEX, NONBLACK and UNEMR83 have significant effects; note that the SOUTH effect on employment odds (or logits) is *positive*, but not significant. The logit equation for Y_2, which includes a lagged effect for the *levels* of Y_1, indicates that the same covariate effects are significant and that some effects have become stronger and some weaker. The lagged effect of Y_1 ($\hat{\beta}_1 = 2.773$) is just the partial log-odds ratio between Y_2 and Y_1, so accounting for heterogeneity in the sample with respect to these covariates still leaves a great deal of time-dependence to be explained. Finally, the logit equation for Y_3 shows some differences in the effects of covariates (NONBLACK is no longer significant, SOUTH becomes significant). The effect of Y_1 (indicating second-order dependence) is cut approximately in half, but the first-order lagged effect is still substantial ($\hat{\beta}_2 = 1.860$). We could consider various possible explanations for the time dependence by either introducing more covariates or by

TABLE 12.10. Time-Sequence Logit Models Involving Fixed Covariates: Three Waves, Dichotomous Version of Labor Force Status

Predictor[a]	Parameter estimates and standard errors for equation predicting		
	Logit(Y_1)	Logit(Y_2)	Logit(Y_3)
Constant	.313	−.539	−.720
	(.287)	(.365)	(.370)
	Lagged effects		
Y_1	—	2.773*	1.395*
		(.127)	(.142)
Y_2	—	—	1.860*
			(.142)
	Covariate effects		
SEX	.817*	.971*	.922*
	(.107)	(.134)	(.150)
NONBLACK	.748*	.356*	.150
	(.119)	(.150)	(.153)
RU83	.261	−.077	.025
	(.136)	(.172)	(.173)
SOUTH	.181	.173	.308*
	(.112)	(.139)	(.143)
UNEMR83	−.042*	−.063*	−.073*
	(.016)	(.020)	(.020)

[a] Definitions of predictors are as follows: SEX (1 = male, 0 = female), NONBLACK (1 = nonblack, 0 = black), RU83 (1 = urban in 1983, 0 = rural in 1983), SOUTH (1 = residence in South in 1983, 0 = non-South), UNEMR83 (unemployment rate for "local labor market" in 1983). Sample size was $N = 2084$. Y_1, Y_2, and Y_3 are labor force status, dichotomous version, for 1983, 1984, and 1985, respectively. Asterisks indicate statistical significant (two-tailed test) at the .05 level.

introducing time-varying covariates (such as the unemployment rate in local labor markets at time 2 in addition to time 1). There are many possibilities and quite a number of ways that competing explanations for the time dependence can be entertained.

We next consider the ordinal logit model of (16), again estimated as a time-nested sequence of individual logit equations. Results appear in Table 12.11 for each of the three equations that make up this model. The constant terms ($\hat{\alpha}_i$) are of little interest in themselves. Recalling the fact

TABLE 12.11. Time-Sequence Logit Models Involving Fixed Covariates: Three Waves, Fourfold Version of Labor Force Status (ordinal logit model)

Predictor[a]	Parameter estimates and standard errors for equation predicting		
	Logit(Y_1)	Logit(Y_2)	Logit(Y_3)
Constant 1	1.153*	−.378	−.627*
	(.240)	(.275)	(.279)
Constant 2	.566*	−1.173*	−1.403*
	(.239)	(.277)	(.281)
Constant 3	−.529*	−2.589*	−2.874*
	(.239)	(.283)	(.288)
	Lagged effects		
Y_1 (linear)	—	1.194*	.489*
		(.047)	(.051)
Y_2 (linear)	—	—	.829*
			(.052)
	Covariate effects		
SEX	.837*	.746*	.789*
	(.085)	(.096)	(.098)
NONBLACK	.542*	.179	.055
	(.101)	(.112)	(.114)
RU83	.128	.045	.160
	(.114)	(.125)	(.126)
SOUTH	.264*	.252*	.176
	(.092)	(.102)	(.102)
UNEMR83	−.046*	−.038*	−.038*
	(.013)	(.143)	(.014)

[a] See notes to Table 12.10 for description of covariates. The constant terms are intercepts in the three logit equations specified by cutting the fourfold response in the three possible ways (see text). The linear terms for Y_1 and Y_2 are used in place of dummy variables, i.e., scores 0, 1, 2 and 3 are assigned to the categories. Most of the lagged effects can be captured by the linear terms. Asterisks denote estimates that are statistically significant at the .05 level (two-tailed test).

that the specified response variable in each equation is the generalized logit contrasting higher versus lower levels of Y_t, with the same covariate effects posited for each contrast (three in all for a four-category response), the parameter estimates have interpretations similar to the interpretations above for the dichotomous logit models. SEX (male = 1) is positively and strongly related to higher levels of labor force activity; NONBLACK is significant only for the first wave, with the effect diminishing over time once we control for previous labor force states; RU83 is insignificant in all waves; the SOUTH effect is strong in two waves but not in the third. Finally, the contextual variable, UNEMR83, is relatively strong in each equation, reflecting the persistent effect of prior (1983) labor market conditions on subsequent labor force outcomes. Comparing results between Tables 12.10 and 12.11, we see that many of the same kinds of effects are present for the dichotomous and the fourfold version of labor force status.

Finally, the time dependence or lagged effects are substantial. For the second wave, the linear effect of Y_1 on the ordinal logit is substantial ($\hat{\beta}_1 = 1.194$). The second-order lag in the equation for the third wave is smaller but still significant; the effect of time 1 status on time 3 status is about one half of the effect of time 1 status on time 2 status. In short, the results indicate a kind of persistent "state dependence" that cannot be accounted for by sample heterogeneity with respect to the set of predictors included in these equations.

6.3. Covariate Effects on Symmetry, Persistence, and Association

For the 2×2 table cross-classifying Y_1 and Y_2 (both dichotomous), the model in (4) indicated how a special design matrix could be used to code three effects of special interest in studying change. These effects can themselves be analyzed by introducing covariates. We illustrate with two possibilities, the first where all covariates are discrete and the second where some of the covariates might be continuous.

It is sufficient to illustrate these by considering the log-frequency model for the table cross-classifying Y_1, Y_2, and a single dichotomous covariate B. With symmetry, persistence, and association parameters defined as in (3.3), the model with main effects on these three terms is

$$\log(F_{ijk}) = \lambda + \lambda_1^B X_k^B + \lambda_S Z_{ij}^S + \lambda_P Z_{ij}^P + \lambda_A (Z_{1i} Z_{2j})$$

$$+ \lambda_{SB}(Z_{ij}^S X_k^B) + \lambda_{PB}(Z_{ij} X_k^B) + \lambda_{AB}(Z_{1i} Z_{2j} X_k^B). \quad (17)$$

Here X_k^B is a dummy variable or a similar variable that contrasts levels of the covariate B. This model is actually saturated for a $2 \times 2 \times 2$ table cross-classifying the two measures of Y and the single dichotomous covariate. Note that the model gives more relevant information about the change processes, as before, and how these processes might depend on the covariate B. The model is generalized for any number of covariates by including all main effects and interactions among the covariates, plus the symmetry, persistence, and association effects, plus any interactions of the covariates with these effects.

To illustrate the application of such models, the 1984 and 1985 dichotomous versions of Y were used along with dichotomous predictors SEX and NONBLACK. Results from the saturated model and a reduced model appear in Table 12.12. The parameter values pertaining to the constant and the main effects of the covariates have been omitted. Results indicate that the association between Y_1 and Y_2 (corresponding to the λ_A parameter and interactions) is substantial, but that it does not depend on sex or color, a surprising result. In the reduced model, we find $\hat{\lambda}_A = .636$; $4(.636) = 2.544$ is the log-odds-ratio between Y_1 and Y_2, adjusting for the other effects, corresponding to an odds-ratio of 12.73. We also find that the symmetry effects are nil, for each sex-color group.

TABLE 12.12. Covariate Effects on Symmetry, Persistence, and Association Parameters: 1984 by 1985 Dichotomous Labor Force Status

	Model			
	Saturated[b]		Reduced[c]	
Effects[a]	$\hat{\lambda}$	$\hat{\lambda}/s(\hat{\lambda})$	$\hat{\lambda}$	$\hat{\lambda}/s(\hat{\lambda})$
Symmetry—Main	.078	1.26	—	
By SEX	−.035	−.57	—	
By NONBLACK	.063	1.03	—	
By SEX/NONBLACK	.059	.97	—	
Persistence—Main	−.731	−18.34	−.719	−19.16
By SEX	.371	9.31	.356	9.76
By NONBLACK	.195	4.88	.208	5.74
By SEX/NONBLACK	−.091	−2.29	−.104	−2.87
Association—Main	.617	16.87	.636	20.55
By SEX	.044	1.20	—	
By NONBLACK	−.032	−.88	—	
By SEX/NONBLACK	.034	.94	—	

[a] Covariates are SEX (1 = male, −1 = female) and NONBLACK (1 = nonblack, −1 = black).
[b] The saturated model has 0 df. Main effects of covariates in the log-frequency model are not reported.
[c] The reduced model has a likelihood-ratio chi-squared value of 6.44 on 7 df.

Note that symmetry in the 2×2 table holds if and only if marginal homogeneity holds; the fact that symmetry holds, approximately, for each sex-color group implies that the marginal distributions of Y_1 and Y_2 are essentially homogeneous for each sex-color group. Finally, there are strong persistence effects, which can be interpreted in the following way. For nonblack males the effect is $-.719 +$ staying in state 1 (not in labor force) versus staying in state 2 (in the labor force). The odds of staying in state 1 versus staying in state 2 is thus .60 for nonblack males. Using the parameter values in this way for the other groups gives the corresponding odds of .22 (nonblack females), .39 (black males), and .06 (black females). The sex-color groups have very different levels of relative persistence as measured by these quantities.

The form of the contrast vectors used to code symmetry, persistence, and association for the 2×2 table leads to a set of three logit models that can be estimated separately. An important property of the respecified design is that the *estimates* of symmetry effects will be uncorrelated with the *estimates* of the persistence effects, although both will be correlated with *estimates* of association effects. We can introduce discrete and/or continuous covariates to explain these effects by considering the following set of individual-level logit models:

$$\Phi_h^S = \log\left[\frac{P(Y_1 = 1, Y_2 = 2)}{P(Y_1 = 2, Y_2 = 1)}\right] = \alpha + \gamma_S' \mathbf{x}_h, \tag{18a}$$

$$\Phi_h^P = \log\left[\frac{P(Y_1 = Y_2 = 1)}{P(Y_1 = Y_2 = 2)}\right] = \alpha^* + \gamma_P' \mathbf{x}_h, \tag{18b}$$

$$\Phi_{2h} - \Phi_{1h} = \log\left[\frac{P(Y_1 = Y_2 = 1)P(Y_1 = Y_2 = 2)}{P(Y_1 = 1, Y_2 = 2)P(Y_1 = 2, Y_2 = 1)}\right]$$
$$= \alpha^{**} + \gamma_A' \mathbf{x}_h. \tag{18c}$$

In these equations, Φ_h^S and Φ_h^P are the logits for symmetry and persistance, respectively, and Φ_{1h} and Φ_{2h} are the logits on Y_1 and Y_2, respectively. The covariates are included in the $\mathbf{x}_h$ vector, which is the value of the covariate set for the hth individual. Estimates of equation (18a), an ordinary logit model, are based on different cases than the estimates of equation (18b); the first equation models only stayers, whereas the second equation models only movers. Because of this, the estimates from the two equations are independent of each other. The first two equations can thus be estimated and tested independently. The third equation is the same as a simple logit model for Y_2 including the levels of Y_1 on the right-hand side along with the association parameters and the interactions of $\mathbf{x}_h$ with Y_1. This equation can thus be estimated in a straightforward way as a logit model that includes the lagged effect of Y_1, covariate effects on the logit, and interactions of Y_1 and the covariates. (The γ_A parameters would appear as the coefficients of interactions involving $\mathbf{x}_h$ and Y_1.) These estimates will, however, be correlated with estimates from the first two equations. In sum, symmetry, persistence, and association effects can be modelled in straightforward ways using discrete or continuous covariates by recasting each component as a standard logit model.

The apparatus for dealing with multiple waves of a dichotomous response or with polytomous responses (observed in two or more waves) is cumbersome to develop. To save space, specifics will not be included here. However, we note that the generalizations required build on the models considered previously in this chapter that were formulated for cases without covariates.

6.4. Covariate Effects on Markov Models

It was demonstrated earlier that Markov models can be expressed as logit or multinomial logit models. In fact, the time-sequence logit models considered earlier in this section are Markov models when the covariate effects are deleted. For example, a first-order Markov model contains first-order lags, a second-order Markov model contains first- and second-order lags, etc. To introduce covariate effects in these models, the apparatus in Sections 6.1 and 6.2 can be applied directly. Special models with persistence effects can be considered easily with either the log-frequency or the logit versions of these models.

7. Discussion

In surveying how standard contingency table models can be adapted for the analysis of change, we have used the tactic of presenting a variety of relevant models or parameterizations. These models are suited for data structures that arise in a multiwave panel study with a small or moderate number of waves. For the case where there are many waves, or many repeated measurements, the conventional contingency table framework utilized here may not be entirely appropriate, although this framework at least suggests possibilities. Recent work on conditional and/or quasi-likelihood methods appears to be more suited for the analysis of many repeated measures. The January 1988 issue of *Statistics in Medicine,* which is devoted to the analysis of repeated categorical measurements, contains both theoretical and applied material on these additional topics as well as references to the literature. (Indeed, the bibliography for this chapter is shortened because of the availability of this key source.)

Some of the models considered here were developed rather completely and demonstrated with empirical examples. For other models, however, we hope that the treatment provided here will at least suggest the appropriate generalizations. For example, there are several ways that the special design matrices for 2×2 and $2 \times 2 \times 2$ tables can be modified for $I \times I$ and $I \times I \times I$ tables and used with covariate effects, even though this exposition does not go into details. Still another area for generalization is the consideration of multivariate response variables, such as the case where two or more dichotomous or polytomous response variables are available at each occasion of measurement. In fact, such models are closely related to methods that would be used to incorporate time-varying covariates. Early work on this subject by Goodman (1978) can be generalized for multivariate settings using many of the ideas presented here, but we leave the specifics to some later report.

References

Agresti, A. (1984). *Analysis of ordinal categorical data.* New York: Wiley.

Becker, M. P., and Clogg, C. C. (1988). "A note on approximating correlations from odds ratios," *Sociological Methods and Research.* **16,** 407–424.

Bishop, Y. M. M., Fienberg, S. E., and Holland, P. W. (1975). *Discrete multivariate analysis: theory and practice.* Cambridge, Mass.: MIT Press.

Clogg, C. C. (1984). "Some statistical models for analyzing why surveys disagree." In C. F. Turner and E. Martin (Eds.), *Surveying subjective phenomena,* Vol. 2 (pp. 319–366). New York: Russell Sage Foundation.

Clogg, C. C., and Eliason, S. R. (1987). "Some common problems in log-linear analysis," *Sociological Methods and Research.* **16:** 8–44. Reprinted in 1988 in J. Scott Long (Ed.), *Common problems/proper solutions: avoiding error in quantitative research* (pp. 226–257). Newbury Park, Calif.: Sage.

Clogg, C. C., and Shockey, J. S. (1988). "Multivariate analysis of discrete data." In J. R. Nesselroade and R. B. Cattell (Eds.), *Handbook of multivariate experimental psychology* (pp. 337–365). New York: Plenum.

Fienberg, S. E. (1980). *The analysis of cross-classified categorical data.* Cambridge, Mass.: MIT Press.

Forthofer, R. N. and Lehnen, R. G. (1981). *Public program analysis: a new categorical data approach*. Belmont, Calif.: Wadsworth.

Goodman, L. A. (1968). "The analysis of cross-classified data: independence, quasi-independence, and interactions in contingency tables with or without missing entries," *Journal of the American Statistical Association*. **63**, 1091–1131.

Goodman, L. A. (1973). "The analysis of multidimensional contingency tables when some of the variables are posterior to others: a modified path analysis approach," *Biometrika*. **60**, 179–192.

Goodman, L. A. (1978). *Analyzing qualitative/categorical data*. Cambridge, Mass.: Abt Books.

Goodman, L. A. (1984). *The analysis of cross-classified data having ordered categories*. Cambridge, Mass.: Harvard University Press.

Haberman, S. J. (1978). *Analysis of qualitative data, Vol. 1. Introductory topics*. New York: Academic Press.

Haberman, S. J. (1979). *Analysis of qualitative data. Vol. 2. New developments*. New York: Academic Press.

Hout, M., Duncan, O. D., and Sobel, M. E. (1987). "Association and heterogeneity: structural models of similarities and differences." in C. C. Clogg (Ed.), *Sociological methodology 1987* (pp. 145–184). Washington, D.C.: American Sociological Association.

Knoke, D., and Burke, P. J. (1980). *Log-linear models*. Beverly Hills, Calif.: Sage.

McCullagh, P. (1980). "Regression models for ordinal data (with discussion)," *Journal of the Royal Statistical Society, Ser. B*. **42**, 109–42.

Reiser, M., Wallace, M., and Schuessler, K. (1986). "Direction-of-wording effects in dichotomous social life feeling items." In N. B. Tuma (Ed.), *Sociological methodology 1986* (pp. 1–25). Washington, D.C.: American Sociological Association.

Sobel, M. (1988). "Some models for multi-way contingency tables with a one-to-one correspondence among categories." In C. C. Clogg (ed.), *Sociological methodology 1988*. Washington, D.C.: American Sociological Association.

Sobel, M. E., Hout, M., and Duncan, O. D. (1985). "Exchange, structure, and symmetry in occupational mobility," *American Journal of Sociology*. **91**, 359–372.

Ware, J. H., and Lipsitz, S. (1988). "Issues in the analysis of repeated categorical outcomes," *Statistics in Medicine*. **7**, 95–107.

Chapter 13

Testing Developmental Models Using Latent Class Analysis*

DAVID RINDSKOPF

City University of New York Graduate Center
New York, NY

Abstract

Psychology in general has moved away from using typologies and toward using continua to conceptualize dimensions of behavior, but certain aspects of learning and development are still fruitfully considered in typological terms. One reason for the abandonment of typologies was that a small number of types seemed unable to explain the enormous diversity of behavior. The development of statistical models such as latent class analysis has allowed theories involving types to be tested, and the allowance for errors of measurement explains how a small number of types can result in a large number of observed patterns of behavior. This chapter demonstrates the application of latent class analysis in several areas of interest to developmental psychologists.

Many years ago, psychological theories based on typologies began giving way to theories based on traits; discrete categories were considered crude approximations to continuous traits. If only measuring instruments were

* The author wishes to thank Geoffrey Saxe for providing one of the data sets discussed here. This Chapter is based on a paper presented at a Social Science Research Council Workshop in New York City, in February, 1985. The first part of this chapter is adapted by permission of Academic Press, Rindskopf, D. *Developmental Review* **7**, 66–85 (1987).

precise enough, it was thought, any characteristics could be measured quantitatively.

But many theories, particularly in developmental psychology, have continued to be more closely aligned with type than with trait models. Children can either conserve volume or they cannot; Piagetian theory does not measure conservation on a continuous scale. Many other examples of a similar kind could be found; they have in common the characteristic that people are hypothesized to be in categories, instead of on a continuous scale. This chapter describes the theory and application to developmental psychology of a method for the statistical analysis of data based on the existence of classes or categories of people. The method, called *latent class analysis* (or *latent structure analysis*), was developed over 30 years ago, but has only become easy to implement with the development of high-speed computers.

The simplest latent class models consider only two classes (types) of people, whereas more complicated models consider more than one trait; there are also models where traits form a hierarchy of development. Section 1 describes the conceptual background necessary for understanding the basic model. Later sections show some applications to data sets, and show the relationship of latent class analysis to other statistical methods. The final section contains instructions for fitting models using function maximizers or minimizers. This allows a great deal of flexibility in the variety of models that can be fit.

1. Conceptual Overview of Latent Class Analysis

To start with the simplest possible case, suppose that a theory hypothesizes that there are two kinds of people: those who have some skill, characteristic, or trait; and those who do not. This might be those who have and have not achieved formal operations, for example. If four items designed to measure this characteristic were presented to a group of people, and if each item were a perfect measure of the characteristic, then only two response patterns would be observed. Those who have acquired formal operations would get all of the items correct (which can be represented by the pattern 1111), and those who have not would get every item wrong (represented as 0000). Unfortunately, items are not all perfect, nor are peoples' responses to items always perfect reflections of their skill or ability. We would expect, therefore, that some people who have acquired formal operations would miss one or more items due to carelessness, fatigue, misunderstanding the wording of the item, and so on. We also might expect that those who have not acquired formal operations might get one or more items right, perhaps by guessing or cheating, or because the items might allow a correct response to be deduced by means other than those the items are supposed to test. We would then see other patterns than the perfect patterns (0000 and 1111); with errors of measurement any of the 16 response patterns is possible. Historically, this discrepancy between theory and responding has been labeled the "competence performance" distinction (Zimmerman and Whitehurst, 1979).

When patterns other than the perfect patterns are observed, a theorist has two options. One option is to abandon the theory that there are two kinds of people, and adopt a theory that people fall on a continuous scale. In this case, counting the number of correct items will place each person on the scale. The other option is to retain the theory that there are two kinds of people, but presume that error processes such as those described earlier might account for the presence of the other observed response patterns. If there are errors in responses to items, then the actual class to which a person belongs is not certain; it is not directly observed, but is inferred through the pattern of right and wrong answers, which contain errors. The "real" class to which people belong is called their *latent* (or unobserved) *class,* from which the name latent class analysis for the statistical method of testing such models is derived.

Latent class analysis would not be very useful if it merely speculated that the occurrence of these "imperfect" response patterns were due to error. Its utility comes from the mathematical model that makes predictions about the way these responses will be distributed. If a particular model is correct, then the number of people displaying each possible response pattern will have a predictable distribution. If this distribution does not occur, then there is evidence that the model is wrong. In other words, latent class models are falsifiable.

2. Procedural Outline of Latent Class Analysis

In this section, the basic steps in latent class analysis are described conceptually. For details on the technical aspects of latent class analysis, see Goodman (1974a,b) and Haberman (1974, 1977, 1979).

The first step in doing a latent class analysis is to specify the model. In the preceding example, there are two latent classes, and there are four observed variables, all of which are dichotomous. In general, both latent and observed variables must be categorical, but need not be dichotomous. In each sample of people, a certain proportion will be in each latent class. The probabilities of being in each of the classes are called the unconditional latent class probabilities. These are represented symbolically as π_t^X, where π indicates that we are representing a proportion, X is the latent variable, and t is the level (category) of X (i.e., the latent class) to which we are referring.

In the foregoing example, for those in the latent class that has attained formal operations, we can consider the probability of answering a particular item correctly. Presumably this would be rather high, but the probabilities may be different for each item; that is, some items are harder than others. There are, therefore, four conditional probabilities of answering items correctly given membership in the latent class that has attained formal operations. Similarly, for those who have not attained formal operations, there are conditional probabilities of answering each of the four questions correctly. These are presumably much lower than the corresponding conditional probabilities for those who have attained formal operations. If we let A, B, C, and D represent the observed variables, then (for example) $\pi_{it}^{\bar{A}X}$ will represent the probability of observing response i on variable A, given that a person is in class t. For all the examples in this chapter, the observed variables are dichotomous, so $i = 1$ or 2. The unconditional and conditional probabilities considered together are the parameters (i.e., unknown constants that must be estimated) in the statistical model.

Once the model is specified, the parameters must be estimated. Usually this is done using the method of maximum likelihood. Two low-cost computer programs are available for doing the calculations required in latent class analysis. One, called MLLSA (maximum likelihood latent structure analysis) was written by Clifford Clogg (1977). The source code in FORTRAN for another program, called LAT, is listed in an appendix of Haberman (1979). (Although both of these programs are written in FORTRAN, once they have been installed on a computer system, no knowledge of FORTRAN is necessary except for simple formatting statements in the LAT program. Both programs can be easily installed on either mainframes or microcomputers). Those interested in pursuing technical details should consult Goodman (1974a,b) or Haberman (1979) and the references they contain.

One general point remains about the specification of the statistical model. Although the observed variables are certainly related, the latent class model assumes that they are unrelated *within* latent classes. That is, in any latent class, errors on each item are independent of errors on other items. This is similar to independence assumptions in other statistical models, such as factor analysis, and merely asserts in a statistical way that the latent classes explain all of the relationships among the observed variables. Because of the conditional independence assumption, the joint probabilities for being in a particular latent class *and* giving particular responses on the observed variables can be calculated as $\pi_{ijklt}^{ABCDX} = \pi_t^X \pi_{it}^{\bar{A}X} \pi_{jt}^{\bar{B}X} \pi_{kt}^{\bar{C}X} \pi_{lt}^{\bar{D}X}$. These can be thought of as the frequencies in a table that is not directly observed (because we never know which class someone is in), and is therefore called the *indirectly observed table.* Summing these values over classes gives the expected frequencies in the *directly observed table.*

After the parameter estimates are obtained, the next important step is to test the fit of the model. That is, a statistical test is done to see whether the observed data are consistent with the statistical model which is being hypothesized. This is done using a chi-square test. The expected frequencies for this test are obtained in a simple manner from the parameter estimates. These expected frequencies are compared with the observed frequencies using either a Pearson or likelihood ratio chi-square statistic. Although the Pearson statistic is the more familiar one, there are reasons for preferring to use the likelihood ratio statistic in most situations, as will be seen when the comparison of different latent class models is discussed. The likelihood ratio chi-square is computed using the formula $2\sum_i O_i \ln(O_i/E_i)$, where O_i represents the observed frequency in cell i, and E_i represents the expected frequency in cell i.

If the model that is being tested is correct, then the expected frequencies should be close to the observed frequencies, and the chi-square statistic will be small. If the model is false, then the expected frequencies will not be close to the observed frequencies, and the chi-square value will be large. Large chi-square values will lead to the rejection of models as being implausible, since they are unlikely to have generated the observed data, and small chi-square values will not lead to rejection of the model. In order to evaluate the chi-square statistic, the critical value from the chi-square distribution is needed. To do this, the degrees of freedom must be counted.

The degrees of freedom for testing a latent class model are the number of independent observed cell proportions minus the number of independent parameters estimated. The number of independent proportions observed is one less than the number of observed cells in the cross-tabulated data, since the proportions must sum to 1. In the previous example there are four dichotomous variables, and 16 possible response patterns in the frequency distribution. The number of independent proportions is therefore $16 - 1 = 15$.

There is one independent unconditional probability in the model, since once that value is known, the other is calculated by subtraction from 1. In general, if there are k classes, then there are $k - 1$ independent unconditional probabilities, because they must sum to 1. There are four independent conditional probabilities in each latent class, corresponding to the probability of getting each item correct given membership in a latent class. The conditional probabilities of missing items are merely 1 minus the probabilities of getting the corresponding items correct. The total number of independent parameters estimated in this model is therefore $1 + 4 + 4 = 9$, and the degrees of freedom for testing this model are $15 - 9 = 6$.

3. Example: Conservation of Weight

Macready and Macready (1974, reported in Dayton and Macready,[1] 1983) tested the ability of 64 children to conserve weight. There were 25 children whose responses on all four items indicated they conserved weight; 25 other children gave responses on all four items indicating they did not conserve; and the remaining 14 children gave mixed responses. The two-class model appears reasonable given the theory; an empirical test of this model results in a likelihood ratio chi-square of 6.742 with 6 degrees of freedom, so the data fit the model reasonably well. The parameter estimates for this model (labeled DM-1) are given in Table 13.1, along with the estimates for several other models for this data set, and the goodness-of-fit statistics for these models. In order to tell which class represents conservers and which represents nonconservers, the conditional probabilities of answering each item correctly are examined. For class 1, these probabilities are low, whereas for class 2 they are high; this establishes class 1 as the class that has not attained conservation of weight, and class 2 as the class that has. Items 3 and 4 are nearly perfect, in the sense that those who can conserve have a probability near 1 of answering them correctly, and those who cannot conserve have probabilities of 0 of answering them correctly. Item 2 is somewhat worse, and item 1 worse yet, but still useful in spite of the fact that only three out of

[1] Dayton and Macready (1983) describe some additional latent class models for this data set.

TABLE 13.1. Analyses of Dayton and Macready Data on conservation

Model	Class	P (class $= i$)	Parameter estimates P (right $\mid$ Class) Item 1	Item 2	Item 3	Item 4
DM-1	1	.42	.00	.07	.00	.00
	2	.58	.76	.84	1.00	.97
DM-2	1	.39	0^a	0^a	0^a	0^a
	2	.35	1^a	1^a	1^a	1^a
	3	.26	.33	.63	.87	.81
DM-3	1	.41	.00	.05	.00	.00
	2	.58	.76	.84	1.00	.97
	3	.02	.00	.84	.00	.00
DM-4	1	.42	.00	.07	.00	.00
	2	.37	1.00	1.00	1.00	.96
	3	.21	.33	.56	1.00	1.00
DM-5	1	.42	.00	.07	.00	.00
	2	.06	.00	.07	1.00	.97
	3	.00	.85	.93	.00	.00
	4	.52	.85	.93	1.00	.97
DM-6	1	.42	.00	.07	.00	.00
	2	.06	.00	.07	1.00	.97
	3	.52	.85	.93	1.00	.97

Model	Goodness-of-fit tests Chi-square (LR)	df^b
DM-1	6.742	10
DM-2	15.086	9
DM-3	6.742	1
DM-4	.001	8
DM-5	.718	8
DM-6	.717	9

[a] Fixed parameter.
[b] Adjusted for parameter estimates on boundary.

four conservers answer it correctly. The fit and parameter estimates for several other models are also listed in Table 13.1; they are discussed in Rindskopf (1987).

4. Hierarchical Models

Theory sometimes dictates that certain skills should be developed in a particular order, so that no one should acquire skill B until skill A has been acquired, and so on. Considering each skill as a dichotomous latent variable, with 4 such skills (which we will call A, B, C, and D), there are 16 combinations of presence and absence of the skills. If the hierarchical theory is correct, only 5 of these 16 types should exist. Using a 0 to represent that a skill has not been acquired, and a 1 to represent that a skill has been acquired, these five patterns can be represented as 0000, 1000, 1100, 1110, and 1111.

Suppose that four items (also called A, B, C, and D) are developed, one of which measures each of the four skills. Because items are not perfect, the conditional probabilities will not all be 0 or 1. However, it should be true that no matter which class people are in, their probability of getting an item correct should be the same as people in other classes who are at the same level with respect to the skill needed for that item. For example, only people in the first latent class listed above lack the skill for item A; therefore people in classes 2, 3, 4, and 5 all have that skill and should have the same probability of getting item A correct. Even though those in the higher classes have more skills, those skills are irrelevant to answering item A correctly. These considerations result in the imposition of equality constraints on the conditional probabilities; their nature will be evident from the parameter estimates.

To demonstrate the hierarchical model and compare it with some other models, a data set kindly provided by Geoffrey Saxe will be used. There are four dichotomous items in this data set; the observed frequencies are listed in Table 13.2. The items are hypothesized to form a hierarchy such as that described before.

The first model fit to these data is the hierarchical model; the goodness-of-fit statistics and parameter estimates are in Table 13.3, where this model is labeled S-1. This model fits the data well. As was true for some models for the first data set, there are more degrees of freedom than would be expected, because three of the parameter estimates are on the boundaries. There are four independent unconditional probabilities

TABLE 13.2. Saxe Data and Results of Fitting a Two-Class Model

Item A	Item B	Item C	Item D	Observed	Expected	Class	P (class | response)
Fail	Fail	Fail	Fail	7	4.62	2	1.00
Pass	Fail	Fail	Fail	16	19.03	2	1.00
Fail	Pass	Fail	Fail	2	2.72	2	1.00
Pass	Pass	Fail	Fail	13	11.59	2	.97
Fail	Fail	Pass	Fail	0	1.05	2	1.00
Pass	Fail	Pass	Fail	6	4.31	2	1.00
Fail	Pass	Pass	Fail	0	.62	2	1.00
Pass	Pass	Pass	Fail	7	7.07	1	.64
Fail	Fail	Fail	Pass	0	0.0	—	—
Pass	Fail	Fail	Pass	0	0.0	—	—
Fail	Pass	Fail	Pass	0	0.0	—	—
Pass	Pass	Fail	Pass	2	2.05	1	1.00
Fail	Fail	Pass	Pass	0	0.0	—	—
Pass	Fail	Pass	Pass	0	0.0	—	—
Fail	Pass	Pass	Pass	0	0.0	—	—
Pass	Pass	Pass	Pass	25	24.95	1	1.00

to estimate, and eight independent conditional probabilities, or 12 parameters in all. This would leave three degrees of freedom to test the model; the computer program indicates six because of the boundary values.

One problem with testing this hierarchical model is that, in general, it is not identified. Rindskopf (1983) discusses ways of making the model identified. One way is to eliminate the second and fourth latent classes, leaving a three-class model in which two classes respond alike to items A and B, and two classes respond alike to items C and D. The results of fitting this model are listed in Table 13.3 as model S-2. As expected, the chi-square value for testing model S-2 is the same as that for testing model S-1. In this model there are two more degrees of freedom, because there are two fewer unconditional probabilities to estimate.

If we were to accept this model, the interpretation would be that there are three kinds of people: those who mastered no skills, those who mastered skills A and B, and those who have mastered all skills. Although this is apparently not consistent with the hierarchical model we started with, in fact the problem is just that there is not the right kind of data to test the full hierarchical theory. The theory may be right, but with

TABLE 13.3.   Analyses of Saxe Data

Model	Class	P (class = i)	Item A	Item B	Item C	Item D
			Parameter estimates			
			P (correct | class)			
S–1	1	.18	.36	.22	.19	.00
	2	.30	1.00	.22	.19	.00
	3	.12	1.00	1.00	.19	.00
	4	.05	1.00	1.00	.93	.00
	5	.35	1.00	1.00	.93	.98
S-2	1	.48	.76	.22	.19	.00
	2	.12	1.00	1.00	.19	.00
	3	.40	1.00	1.00	.93	.98
S-3	1	.08	0*	0*	0*	0*
	2	.32	1*	1*	1*	1*
	3	.61	.94	.51	.28	.05
S-4	1	.41	1.00	1.00	.92	.85
	2	.59	.81	.37	.19	.00

Goodness-of-fit tests

Model	Chi-square (LR)	df
S-1	4.680	6
S-2	4.680	8
S-3	6.634	9
S-4	5.863	9

Note: Degrees of freedom are adjusted for parameters on boundaries.

only one item testing each skill no definitive test of the hierarchical structure is possible.

Even though the model originally suggested has been tested, it is wise to test simpler models also to be sure that no sensible, more parsimonious model fits the data. Two such possibilities in this case are the quasi-independence model and the unrestricted two-class model. The results of fitting these models are presented as models S-3 and S-4 in Table 13.3. Both models fit the data well, and on grounds of parsimony the two-class model would probably be preferred by most theorists.

Table 13.2 shows, for the two-class model, the probability of a person with each response pattern being in the most likely latent class for that response pattern. This assignment to latent classes is easily done using Bayes' theorem, and is included in the output of the MLLSA program.

5. Models for Panel Data

Models discussed in previous sections were all cross sectional; each person was measured only once. When such data are used to test developmental models, it is assumed that people develop skills or abilities in a certain order, and that all people go through each step in the sequence. Measuring people more than one time allows some of these assumptions to be tested. One common, simple longitudinal design is called the *cross-panel design*. People are asked a series of questions at one time, and the same questions again at a second time. In this section, we investigate some models for such a study in which two dichotomous items are asked (measured) at two times.

As an example, consider a data set reported in Landis and Koch (1979). A sample of 354 children were measured at two time points (T1 and T2) on two developmental attributes (A1 and A2), each of which was either present or not each of the two times. We will consider three types of models for these data, along with another related model suggested by the results of one of the three basic analyses.

The simplest model is that there are two latent classes, that both items measure the trait represented by the two classes, and that children remain in the same class at both times of measurement. A second, more complicated, model is that the items both measure the same characteristic (which is either present or not), but that children's status on this characteristic may change from time 1 to time 2. There would then be four latent classes, with the latent characteristic being either present or absent at time 1, and again at time 2. A third model is that the two items measure different (but possibly related) traits, each of which is either present or not, but that there is stability over time in each trait. That is, if a child has the trait measured by item A1 at time 1, then he or she also has the trait at time 2 (and similarly for the absence of the trait, and for the trait measured by A2). As for the second model, this model assumes four latent classes: people either do or do not have the trait measured by item A1, and either do or do not have the trait measured by item A2, giving $2 \times 2 = 4$ combinations.

The second and third models both are specified by making restrictions on the conditional probabilities. In each case, these are equality restrictions: certain pairs of conditional probabilities are constrained to have equal values. Consider, for example, the model that there is one trait, but that status on the trait can be different at time 1 and time 2. There are two classes which have the trait at time 1; they only differ on whether they have the trait at time 2. People in each of these two classes should have the same conditional probability of responding positively at time 1. Similar statements hold for those who do not have the trait at time 1, and about the conditional probabilities at time 1 for those who do (and do not) have the trait at time 2. When examining the parameter estimates for models 2 and 3, note the pairs of conditional probabilities that have equal values.

The results of fitting these models to the Landis and Koch data are reported in Table 13.4. The two-class model (labeled LK-1 in the table) has a large chi-square value, and is rejected as being improbable for these data. Model LK-2 specifies that there are two traits, one measured by item A1, the other by A2, and that these traits are either present or absent. The combinations of presence and absence of the two traits are indicated by + and − signs in the right margin of the table for this model. The model does not fit the data well, but the two conditional probabilities

TABLE 13.4.   Analyses of Landis and Koch Data

Model	Class	P (class = i)	A1T1	A2T1	A1T2	A2T2		
				Parameter estimates P (+ | class)				
LK-1	1	.50	.78	.89	1.00	1.00		
	2	.50	.02	.00	.46	.58		
							A1	A2
LK-2	1	.00	.78	.00	1.00	.58	+	−
	2	.50	.02	.00	.46	.58	−	−
	3	.50	.78	.89	1.00	1.00	+	+
	4	.00	.02	.89	.46	1.00	−	+
							T1	T2
LK-3	1	.36	.03	.00	.24	.41	−	−
	2	.00	.79	.90	.24	.41	+	−
	3	.15	.03	.00	1.00	1.00	−	+
	4	.49	.79	.90	1.00	1.00	+	+

Goodness-of-fit tests

Model	Chi-square (LR)	df
LK-1	33.908	9
LK-2	33.914	6
LK-3	6.635	5

Note: Parameter estimates on boundaries were ignored in computations of degrees of freedom.

of zero indicate that caution is needed. Several different sets of starting values were tried to see if the chi-square could be decreased, all without success; in each case, the solution reported in the table was obtained. This model is evidently not supported by the data.

The third model, labeled LK-3 in Table 13.4, is that there is one dichotomous trait measured by both variables, and that people may change status on that trait between time 1 and time 2. The meaning of each of the four latent classes is indicated by the columns in the right margin labeled T1 and T2, and in which + and − signs are used to represent presence and absence of the trait at each time. This model fits the data very well, as indicated by the small chi-square value.

Examination of the parameter estimates reveals that one of the latent classes has no people; that is, the unconditional probability is zero. This is that class which has the trait at time 1, but not at time 2. We conclude that "regression" (from possession of the trait to lack of it) does not occur. To properly finish off this analysis, we would fit the three-class model that omits the second class from this model. These results are not reported, as the goodness-of-fit and parameter estimates are the same as those reported for this model, except, of course, that class 2 is not included.

6. Multiple-Group Models

The models discussed so far involved one group of people, whether measured at one time (cross sectional data) or at more than one time (longitudinal data). In this section we consider the case where more than one group of subjects is measured. The groups might consist of subjects of different ages, racial or ethnic groups, geographic areas, sexes, and so on.

Models for multiple groups will generally be very similar to models for a single group, except that the latent classes will be presumed to exist in each group to possibly different extents. That is, one group may have a larger proportion of its members in a latent class than does another group. In the extreme case, some groups may not have any of their members in certain latent classes, thus further restricting the model. Other types of restrictions may also be tested for multiple-group models: it would usually not be unreasonable to assume that in each group, the conditional probabilities might be equal even though the proportions of people in each class might vary across groups. In other cases, it might be expected that the proportions in each class might not vary across groups (i.e., that a distribution of a trait is invariant across populations). For a test of this, we would have to obtain a random sample from each population since the unconditional probabilities are affected by the sampling scheme.

As an example of a multiple-group data set, we present some analyses of the responses of a group of high school students to the same question asked at two different times (the 9th and 12th grades). These data were analyzed using different procedures by Marascuilo and Serlin (1979). The 1652 students were asked whether the following statement was true or false: "The most important qualities of a husband are determination and ambition." Five racial/ethnic groups were represented in the study (Asian, Black, Chicano, (American) Indian, and White).

Although there are a number of plausible latent class models for measurements made at two times (as described in the previous section), the analysis of this data set is severely limited because only one question was asked at each time point. Specifically, we cannot test whether there was a change in attitude over time without making some strong assumptions about restrictions in certain parts of the model. We will consider, out of necessity rather than choice, models that assume stability of the trait from one time to the next, so that the items represent independent assessments of a stable attitude.

The first model tested assumes that there are two latent classes. Because of the nature of the item, these classes might be labeled Traditional (or Conservative) and Nontraditional attitude toward sex roles. The results of fitting the unrestricted two class model are presented in Table 13.5, where the model is labeled MS-1. This model fits the data well, as indicated by the small chi-square value relative to the degrees of freedom.

The conditional probabilities of responding "true" to the question at each time are in the last two columns of the table. Latent class 1 has a lower probability of answering "true" than does latent class 2, which means that class 1 is the Nontraditional class and class 2 is the Traditional class. There is a tendency toward polarization over time: the Traditional class is more likely to answer in a traditional way at time 2, and the Nontraditional class is more likely to answer in a nontraditional way at time 2. These tendencies could be tested by fitting a model with restrictions on the conditional probabilities.

The unconditional probabilities in this table are presented in a different fashion from previous models, because of the presence of several groups.

TABLE 13.5.   Analyses of Marascuilo and Serlin Data

Model	Class	Asian	Black	Chicano	Indian	White	Time 1	Time 2
				Parameter estimates P (class = i | group)			P (true | class)	
MS-1	1	.49	.28	.24	.20	.55	.31	.01
	2	.51	.72	.76	.80	.45	.70	.82
MS-2	1	.32	.32	.32	.32	.32	.38	.20
	2	.68	.68	.68	.68	.68	.83	1.00
MS-3	1	.54	.24	.24	.24	.54	.31	.01
	2	.46	.76	.76	.76	.46	.70	.82
MS-4	1	.50	.28	.24	.19	.56	.30	.04
	2	.11	.22	.14	.35	.14	.30	.80
	3	.39	.49	.61	.46	.30	.89	.80

Goodness-of-fit test

Model	Chi-square (LR)	df
MS-1	8.560	6
MS-2	98.732	10
MS-3	11.840	9
MS-4	2.518	1

Here, the unconditional probabilities are presented separately for each group; that is, they are really conditional on group membership. This is much more informative in the multiple-group case than simple unconditional probabilities. In examining the unconditional probabilities, it is obvious that there are differences among the groups in the proportion of people in each group who have Traditional rather than Nontraditional values. The Asians and Whites are split approximately in half, whereas the Blacks, Chicanos, and Indians are heavily Traditional.

To further refine the model, we test whether it is plausible to assume that all of the groups have the same distribution in the latent classes. From our examination of the unconditional probabilities, this would not seem to be so: the probability of being in class 1 varies from .20 for Indians to .55 for Whites. This is confirmed in the analysis of model MS-2 reported in Table 13.5; in this model the unconditional probabilities are constrained to be equal for all five groups.

Next, the model that Asians and Whites are distributed similarly, and that Blacks, Chicanos, and Indians are distributed similarly, was tested. This model, labeled MS-3 in Table 13.5, fits the data; additionally, the fit is not significantly worse than the unrestricted model MS-1. Note that the unconditional probabilities of being in class 1 (and similarly for class 2) are the same for Asians and Whites, and are the same for Blacks, Chicanos, and Indians.

Another model was fit to attempt to investigate whether there was change from time 1 to time 2. As indicated above, the most general of such models cannot be tested. A three-class model was tested in which one class was Traditional at both times, another class was Nontraditional at both times, and a third class was different at the two times. (Whether it changed from Traditional to Nontraditional or the reverse is difficult to specify in advance in these models; one must examine the results to find out whch occurred.) The results show that this model, labeled MS-4 in Table 13.5, fit well. This is not surprising since the unrestricted two-class model also fit well; the issue is whether the fit of the three-class is significantly better than that of the two-class model.

Because the two-class model is a special case of the three-class model, we can directly compare their fit. By "special case," we mean that by putting restrictions on the three-class model (in this case, setting the parameters for the third class equal to zero), we get the two-class model. When this happens, the fit of the models can be compared by subtracting the likelihood ratio chi-squares; the difference has a chi-square distribution with degrees of freedom equal to the difference in degrees of freedom for the models. (This is the main reason for preferring likelihood ratio chi-squares, since such comparisons cannot be done using Pearson chi-squares.)

The difference in chi-square values between models MS-1 and MS-4 is 6.04 with 5 degrees of freedom, so the three-class model fits no better than the two-class model. Additionally, the three-class model is not identified, while the other models are all identified. One could try to make further restrictions on this model, such as was done for model MS-3, to increase degrees of freedom and possibly make the model identified.

7. Relationship of Latent Class Analysis to Other Methods of Analysis

Latent class analysis is related to other multivariate analysis methods, which are more familiar to most researchers. It is related to log-linear models, and can be thought of as a log-linear model where one variable, the latent class to which people belong, is unobserved. If we let X represent the latent class variable, and A, B, and C represent observed variables, then the log-linear representation of the latent class model is $[XA][XB][XC]$. That is, X is related individually to A, B, and C; but A, B, and C are conditionally independent given X. This is what is meant by the statement that X "explains the relationship among A, B, and C." Some of the more complicated models discussed earlier are not easily expressed in these terms, because they have either equality restrictions or restrictions that some parameters are zero. Such models do not correspond to the usual hierarchical log-linear models, but can be formulated using the model (or design) matrix approach. This approach to log-linear models is discussed in Rindskopf (1990) and is applied to latent class models by Haberman (1979).

Latent class analysis is also related to other latent variable methods. Although latent class analysis deals with categorical latent and observed variables, factor analysis involves continuous latent and observed variables. Latent trait models have continuous latent variables, but categorical (usually dichotomous) observed variables. In many cases, cluster analysis can be conceptualized as involving a categorical latent variable, but continuous observed variables.

8. Do-It-Yourself Estimation

This section is much more technical than the rest of the chapter. It requires a thorough understanding of the latent class model, familiarity with the general concepts of maximum likelihood estimation, and a knowledge of at least one scientific programming language. Readers without this background may find it helpful to first read Erdfelder's chapter in this volume.

When possible, it is simplest to use existing computer programs for latent class analysis. The easiest to use is MLLSA. It uses an algorithm, called the EM algorithm (Dempster, Laird, and Rubin, 1977), that almost always produces a valid solution. A program by Haberman (1979) allows the specification of a wider variety of models, but is tedious to use and frequently will not converge to a proper solution because the estimation procedure is very sensitive to the starting values provided.

Sometimes, however, because existing programs are not general enough or are not providing proper solutions, you may want to write your own program to do an analysis. The example demonstrated here has equality constraints across groups on the unconditional probabilities of being in each class. This cannot be done in MLLSA, and is difficult using Haberman's LAT program unless you are lucky at guessing start values. Or you may want equality constraints across groups on conditional probabilities. Local dependence models (where responses to some pairs, or more generally sets, of items are not independent within one or more classes) can be done using LAT (but again only if you can find start values that result in convergence). Some (but not all) of these models can be testing using MLLSA if special tricks are used (pairs or groups of items are treated as one item). You may want even more unusual constraints, such as inequality constraints, or making one parameter equal to a fraction of another; all of the above are possible using the methods described here.

Writing your own program is not as difficult as might be thought, if you have access to a routine that will find the maximum or minimum value of a specified function. Perhaps the most widely available such routines for mainframe and minicomputers are found in the subroutines of the International Mathematical and Statistical Libraries (IMSL, 1987). One of these routines is demonstrated in this section. Some of the major statistical packages (such as BMDP and SAS) also have the ability to do these analyses using their nonlinear function fitting routines. On microcomputers, programs such as GAUSS can also be used.

A caution: like all other such programs, these may give wrong answers. The most common problems are failure to converge, convergence to a local maximum, local minimum, or saddle point of the function, or to the wrong value altogether due to insufficient arithmetic precision. Some checks (besides the obvious warning messages that some programs will give) include the following:

1. Check that the derivatives of the likelihood function are zero at the solution point.

2. Run the program several times using different starting values. If the chi-square value is the same each time, you probably have found a reasonable solution point. If the chi-square is the same but the parameter estimates differ, then the model is probably not identified. (This would occur in the model demonstrated here if all five groups were constrained to have the same conditional probabilities. This counterintuitive result—that added restrictions make a model unidentified—also occurs in factor analysis and structural equation modeling; see Rindskopf, 1984 for details.)

3. If the program computes second derivatives, check that all the eigenvalues are nonzero and have the same sign (positive for a minimum, negative for a maximum).

4. Examine all parameter estimates to see if they make sense. For example, all probabilities should be between 0 and 1, and any constraints imposed should be checked to be sure that the solution satisfies the constraints. Parameter estimates on the boundary may indicate problems with the maximization (or minimization), or possibly identification problems.

The IMSL subroutine demonstrated here will find the minimum value of a given function. We want to find the maximum value of a function (the logarithm of the likelihood), so all we need to do is to compute the negative of this function, since minimizing the negative of a function is equivalent to maximizing the function itself. (Draw yourself a picture of a quadratic equation, such as $y = x^2 + 1$, and its negative, $y = -x^2 - 1$, if you are not convinced.)

Two techniques are used in the program whose purpose may not be obvious. The parameters of our model are probabilities, so valid values can range only between 0 and 1. However, function minimizers do not know this, and might find parameter estimates outside the allowable range. To prevent this, we use common transformations (see, e.g., Bock, 1970, or Formann, 1982), the logit and its extension to the multinomial case, and their inverses. For a dichotomous variable with probability p of being in the first category, the logit of p is defined as the natural logarithm of $p/(1 - p)$. Although p can range only between 0 and 1, the logit can take on any real value. It is the logits that will be the parameters in the program we will work with, and we will transform back to probabilities using the inverse function: If l is a logit, $p = \exp(l)/\{1 + \exp(l)\}$. For polytomous variables, define $p_i = \exp(l_i)/\Sigma_j \exp(l_j)$. Here an additional constraint, such as $\Sigma_j l_j = 0$ or $l_k = 0$ for some k is needed.

The other special technique used is an indicator variable, called IND in the program, whose purpose is to select the right parameters depending on the group of subjects. This is further explained in the section describing the subroutine for computing the log-likelihood function.

The analysis is of the Marascuilo and Serlin data discussed earlier; the model fit here is MS-3 in Table 13.4. The input is shown in Table 13.6, a list of the important variables in the program is in Table 13.7, and the output from the program is in Table 13.8. The example is written in the FORTRAN IV programming language. As most programming languages

TABLE 13.6. FORTRAN Program Using IMSL Routine DUMPOL to Estimate
Latent Class Model

```
 1.  //JOB
 2.  //EXEC FORTVCLG,ADDLIB = 'SYS2.IMSL',GOREGN = 300K
 3.  //SYSIN DD *
 4.  C. . . .
 5.  C. . . . PROGRAM INPUT IS L7#LCA10D
 6.  C. . . . USES IMSL ROUTINE DUMPOL (VERSION 10)
 7.  C. . . .    TO CALCULATE ML ESTIMATES OF LATENT CLASS MODEL
 8.  C. . . .    FOR MARASCUILO & SERLIN DATA
 9.  C. . . .
10.          IMPLICIT REAL * 8 (A − H, O − Z)
11.          DIMENSION X(6), XX(6), X1(6),
12.       &FO(4, 5), FE(4, 5), NG(5)
13.          COMMON/CCC/ FO,FE,XX,NG,NTOT
14.          EXTERNAL FUNCT
15.  C. . . .
16.  C. . . . INITIALIZE PARAMETERS (START VALUES)
17.  C. . . . SET NECESSARY CONSTANTS FOR MINIMIZATION ROUTINE
18.  C. . . .
19.          DATA X1/ 1.D0, − .5D0, .2D0, − .8D0, 2 ∗ .2D0/
20.          N = 6
21.          FTOL = 1.0D − 12
22.          MAXFCN = 5000
23.          S = .5D0
24.  C. . . .
25.  C. . . . READ FREQUENCIES, FORM TOTALS
26.  C. . . .
27.          READ(9,121) FO
28.    121   FORMAT(4F4.0)
29.          NTOT = 0
30.          DO 131 J = 1,5
31.            NG(J) = 0
32.            DO 131 I = 1,4
33.              NTOT = NTOT + FO(I,J)
34.    131   NG(J) = NG(J) + FO(I,J)
35.  C. . . .
36.  C. . . . FIND MINIMUM OF FUNCTION (MAXIMIZE LIKELIHOOD)
37.  C. . . .
38.          CALL DUMPOL(FUNCT,N,X1,S,FTOL,MAXFCN,X,F)
39.  C. . . .
40.  C. . . . OUTPUT PARAMETER ESTIMATES
41.  C. . . .
42.          WRITE(6,101)XX(1),XX(6),XX(4),XX(5),XX(2),XX(3)
43.    101   FORMAT(///'       FINAL PARAMETER ESTIMATES'//
44.          *' PROB(CLASS = 1 GIVEN ASIAN, WHITE) = ',F10.6/,
```

(continued)

TABLE 13.6. (*Continued*)

```
45.        *' PROB(CLASS = 1 GIVEN BLACK, HISP, IND) = ', F10.6//,
46.        *' CONDITIONAL PROBABILITIES',//,
47.        *' TIME 1          = ',2F10.5,
48.        */' TIME 2          = ',2F10.5)
49.   C. . . .
50.   C. . . . CALCULATE AND PRINT CHI SQUARE VALUE
51.   C. . . .
52.           CS = 0.
53.           DO 201 I = 1,4
54.           DO 201 J = 1,5
55.    201    CS = CS + 2. * FO(I,J) * DLOG(FO(I,J)/FE(I,J))
56.           WRITE(6,103)CS
57.    103    FORMAT(/'     CHI SQUARE VALUE IS ',F12.6//)
58.   C. . . .
59.   C. . . . PRINT OBSERVED, EXPECTED FREQUENCIES
60.   C. . . .
61.           DO 777 J = 1,5
62.           DO 777 I = 1,4
63.    777    WRITE(6,778) I,J,FO(I,J), FE(I,J)
64.    778    FORMAT(' ',2I5,2F12.5)
65.           STOP
66.           END
67.           SUBROUTINE FUNCT(N,X,F)
68.           IMPLICIT REAL * 8 (A − H,O − Z)
69.           DIMENSION X(N),XX(6),FO(4,5),FE(4,5),PRC(4,2),
70.      &NG(5)
71.           COMMON/CCC/ FO,FE,XX,NG,NTOT
72.   C. . . .
73.           DO 997 I = 1,6
74.           EE = DEXP(X(I))
75.    997    XX(I) = EE/(1.D0 + EE)
76.   C. . . .
77.           F = 0.D0
78.           DO 102 I = 1,4
79.           DO 102 K = 1,5
80.           IND = 0
81.           IF(K.EQ.1.OR.K.EQ.5)IND = 1
82.           PRC(1,1) = (IND * XX(1) + (1 − IND) * XX(6)) * XX(2) * XX(4)
83.           PRC(2,1) = (IND * XX(1) + (1 − IND) * XX(6)) * (1 − XX(2)) * XX(4)
84.           PRC(3,1) = (IND * XX(1) + (1 − IND) * XX(6)) * XX(2) * (1 − XX(4))
85.           PRC(4,1) = (IND * XX(1) + (1 − IND) * XX(6)) * (1 − XX(2)) * (1 − XX(4))
86.           PRC(1,2) = (IND * (1 − XX(1)) + (1 − IND) * (1 − XX(6))) * XX(3) * XX(5)
87.           PRC(2,2) = (IND * (1 − XX(1)) + (1 − IND) * (1 − XX(6))) * (1 − XX(3)) * XX(5)
88.           PRC(3,2) = (IND * (1 − XX(1)) + (1 − IND) * (1 − XX(6))) * XX(3) * (1 − XX(5))
89.           PRC(4,2) = (IND * (1 − XX(1)) + (1 − IND) * (1 − XX(6))) * (1 − XX(3)) * (1 − XX(5))
```

(*continued*)

TABLE 13.6.   (*Continued*)

90.		FE(I,K) = 0.D0			
91.		DO 101 J = 1,2			
92.	101	FE(I,K) = FE(I,K) + PRC(I,J) * NG(K)			
93.	102	F = F − FO(I,K) * DLOG(FE(I,K)/NTOT)			
94.		RETURN			
95.		END			
96.	//GO.FT09F001 DD *				
97.		60	50	22	68
98.		72	30	30	41
99.		62	29	19	25
100.		86	28	47	39
101.		243	208	112	381
102.	/*				

used in scientific programming have the same general structure, if you know any language you can probably understand the concepts involved.

For those who know no FORTRAN, the following may help you read the program: lines starting with a "C" in column 1 are comments. All FORTRAN statements start in column 7. Numbered statements (integers in columns 1 to 5) are labels for marking the end of loops, locations to go to, or supplementary information (such as details about formatting information to be read or written). Variables that represent arrays have parentheses after them, with one or more values that act as subscripts.

TABLE 13.7.   Important Variables in the Program

X1	Start values for the parameters; vector with six values
N	Number of parameters in the model
FO	Observed frequencies; 4×5 matrix
NTOT	Total sample size
NG	Number of people in each group; vector with five values
XX	Parameter estimates (probabilities); vector with six values
X	Parameter estimates (in logit scale); vector with six values
CS	Likelihood ratio chi-square statistic
FE	Expected frequencies under the model; 4×5 matrix
PRC	Joint probabilities of being in a particular class and giving a particular response to the items

TABLE 13.8.   Output from Program for Latent Class Analysis

FINAL PARAMETER ESTIMATES

PROB(CLASS = 1 GIVEN ASIAN, WHITE) = 0.537707
PROB(CLASS = 1 GIVEN BLCK, HISP, IND) = 0.241579

CONDITIONAL PROBABILITIES

TIME 1 = 0.31251   0.69728
TIME 2 = 0.00724   0.81788

CHI SQUARE VALUE IS 11.840049

1	1	60.00000	52.97213
2	1	50.00000	45.10515
3	1	22.00000	23.42648
4	1	68.00000	78.49624
1	2	72.00000	74.92140
2	2	30.00000	29.62773
3	2	30.00000	32.69283
4	2	41.00000	35.75803
1	3	62.00000	58.46468
2	3	29.00000	23.11990
3	3	19.00000	25.51175
4	3	25.00000	27.90367
1	4	86.00000	86.61434
2	4	28.00000	34.25171
3	4	47.00000	37.79518
4	4	39.00000	41.33877
1	5	243.00000	250.02845
2	5	208.00000	212.89633
3	5	112.00000	110.57296
4	5	381.00000	370.50226

For example, XX(4) is the fourth element of a vector called XX, and FE(2, 3) is the element in row 2, column 3 of a matrix called FE.

The general procedure is simple: write a subroutine that will compute the negative of the log-likelihood function (since we have a function minimizer), then write a main routine that (a) initializes all the parameters and declares what arrays will be used, (b) reads the data, (c) calls the function minimizer to find the solution, and (d) prints the solution.

In Table 13.6, all lines are numbered for ease of reference; these numbers are not normally present when the program is run and are not part of the program.

Lines 1–3 are IBM Job Control Language (JCL) for the program, and are written specifically for the computer system used at City University of New York.

Lines 10–14 do "housekeeping" chores, such as declaring all variables to be stored as double-precision numbers, setting the dimensions of arrays, declaring a COMMON area to transmit information to and from the subroutine from the main routine, and declaring the name FUNCT to be an external subroutine.

Line 19 has the start values for the parameters. (These are on the logit scale and can be given any real value, though values between −1 and 1 generally will work best. Do not use values that would make conditional probabilities for different classes all equal.)

Lines 20–23 set values necessary for the function minimizer: the number of parameters in the model, the convergence criterion, the maximum number of evaluations of the function allowed before terminating, and a scaling factor for the dimensions.

Lines 27–34 read in a 2×2 table of frequencies for each of the five groups and form subgroup totals and an overall total. (The data are found in lines 97–101.)

Line 38 calls the IMSL routine DUMPOL to find the minimum of the function.

Line 42 writes the parameter estimates according to the format specification in lines 43–48.

Lines 52–57 calculate and print the likelihood ratio chi-square statistic.

Lines 67–95 contain the subroutine to calculate minus the logarithm of the likelihood function; these are described in more detail next.

Lines 68–71 declare variables used, the size of arrays, and common area.

Lines 73–75 transform the parameters used by the minimizer to probabilities, as described earlier.

Lines 80–89 calculate the joint probabilities (denoted PRC(. , .) in the program) by multiplying the unconditional probability of being in a class by the conditional probabilities of responding in a given way to each item given class membership. Here the indicator variable IND is used to select the right unconditional probabilities. IND is set equal to 1 for Whites and Asians, and to 0 for Blacks, Indians, and Chicanos. When IND equals one, $1 -$ IND is equal to zero; and when IND equals zero, $1 -$ IND equals 1.

XX(1) and XX(6) represent the unconditional probabilities of being in class 1 for Whites and Asians, and for Blacks, Indians, and Chicanos, respectively. Therefore, the expression $IND * XX(1) + (1 - IND) * XX(6)$ is equal to $XX(1)$ for Whites and Asians and to $XX(6)$ for Blacks, Indians, and Chicanos.

By using $XX(1)$ to represent the unconditional probabilities of being in class 1 for Whites and Asians, we have imposed an equality constraint on the unconditional probabilities. Similar equality constraints are imposed on the unconditional probabilities of being in class 1 for Blacks, Indians, and Chicanos by using just one value to represent all three parameters. Note also that the conditional probabilities of agreeing with each item are the same for all five groups. This is another example showing how easy it is to impose equality constraints. (Conditional probabilities of disagreeing with an item are represented as one minus the probabilities of agreeing with the item.)

Inequality constraints may be easily imposed by adding the square of one parameter to another parameter. For example, if $Y = XX(1) + XX(2) * XX(2)$, where $*$ indicates multiplication, then Y must be at least as large as $XX(1)$. Or if we want one parameter to be .8 times another, we just create $Y = .8 * XX(1)$, for example. By using combinations of these techniques, a large variety of equality and inequality constraints can be imposed on the model.

Lines 90–93 calculate the expected frequencies (represented as $FE(.,.)$ in the program) by summing the joint probabilities over latent classes.

Lines 97–101 contain the data for the five groups.
The output, in Table 13.8, is relatively spartan. Since you are designing your own program, you can be either very arcane or very reader-friendly, as your conscience dictates.

9. Summary

Latent class analysis gives the developmental researcher a statistically rigorous way of testing a wide variety of models. It has its limitations: sample sizes needed are larger than found in many developmental studies; only a few items can be used in an analysis if the goodness-of-fit statistics are to be valid; the items must be categorical. Further discussions of these and other limitations are in Rindskopf (1987). In spite of these limitations, the examples here show that latent class analysis should become a standard tool for many researchers. A wide variety of sensible models exists, and computer programs are widely available for testing these models. Latent class analysis allows the formal testing of what might otherwise remain vague theories. With adequate planning, studies can shed light on the reasonableness of typological theories of development.

References

Bock, R. D. (1970). "Estimating multinomial response relations." In R. C. Bose, I. M. Chakravarti, P. C. Mahalanobis, C. R. Rao, and K. J. C. Smith (Eds.), *Essays in probability and statistics.* Chapel Hill, N.C.: University of North Carolina Press.

Clogg, C. C. (1977). *Unrestricted and restricted maximum likelihood latent class analysis: A manual for users.* University Park, Penn: Population Issues Research Office.

Dayton, C. M., and Macready, G. B. (1983). "Latent structure analysis of repeated classifications with dichotomous data," *British Journal of Mathematical and Statistical Psychology.* **36,** 189–201.

Dempster, A. P., Laird, N. M., and Rubin, D. B. (1977). "Maximum likelihood estimation from incomplete data via the EM algorithm (with discussion)," *Journal of the Royal Statistical Society.* **B39,** 1–38.

Formann, A. K. (1982). "Linear logistic latent class analysis," *Biometrical Journal.* **24,** 171–190.

Goodman, L. A. (1974a). "The analysis of qualitative variables when some of the variables are unobservable. Part I. A modified latent structure approach," *American Journal of Sociology.* **79,** 1179–1259.

Goodman, L. A. (1974b). "Exploratory latent structure analysis using both identifiable and unidentifiable models," *Biometrika.* **61,** 215–231.

Haberman, S. J. (1974). "Log-linear models for frequency tables derived by indirect observation. Maximum-likelihood equations," *Annals of Statistics.* **2,** 911–924.

Haberman, S. J. (1977). "Product models for frequency tables involving indirect observation," *Annals of Statistics.* **5,** 1124–1147.

Haberman, S. J. (1979). *Analysis of qualitative data.* Vol. 2.: *New developments.* New York: Academic Press.

IMSL (1987). *User's manual: Math/Library: FORTRAN subroutines for mathematical applications.* Houston: IMSL.

Landis, J. R., and Koch, G. G. (1979). "The analysis of categorical data in longitudinal studies of behavioral development." In J. R. Nesselroade and P. B. Baltes (Eds.), *Longitudinal research in the study of behavior and development.* New York: Academic Press.

Macready, C., and Macready, G. B. (1974). "Conservation of weight in self, others, and objects," *Journal of Experimental Psychology.* **103,** 372–374.

Marascuilo, L. A., and Serlin, R. C. (1979). "Tests and contrasts for comparing change parameters for a multiple sample McNemar data model," *British Journal of Mathematical and Statistical Psychology.* **32,** 105–112.

Rindskopf, D. (1983). "A general framework for using latent class analysis to test hierarchical and nonhierarchical learning models," *Psychometrika.* **48,** 85–97.

Rindskopf, D. (1984). "Structural equation models: Empirical identification, Heywood cases, and related problems," *Sociological Methods and Research.* **13,** 109–119.

Rindskopf, D. (1987). "Using latent class analysis to test developmental models." *Developmental Review.* **7,** 66–85.

Rindskopf, D. (1990). "Nonstandard loglinear models," Psychological Bulletin, in press.

Zimmerman, B. J., and Whitehurst, G. J. (1989). "Structure and function: a comparison of two views of the development of language and cognition." In G. J. Whitehurst and B. J. Zimmerman (eds.), *Functions of language and cognition.* New York: Academic Press.

Deterministic
Developmental
Hypotheses, Probabilistic
Rules of Manifestation,
and the Analysis
of Finite Mixture
Distributions*

Chapter 14

EDGAR ERDFELDER

Department of Psychology
University of Bonn
Bonn
Federal Republic of Germany

Abstract

The empirical evaluation of deterministic developmental theories using qualitative theoretical terms (for example, stage theories of human development) is difficult if exact empirical indicator variables for theoretical terms are missing. The possibility of linking theoretical and empirical variables by means of probabilistic rules of manifestation is investigated. This approach leads to a broad class of statistical models, namely, finite mixture distributions. The adequate modeling of developmental theories in the framework of finite mixtures, identifiability, parameter estimation,

* This chapter is a considerably revised version of a paper entitled "Deterministische allgemeinpsychologische Hypothesen, probabilistische Manifestationsregeln und die Analyse latenter Klassen," which the author presented at the 34th Kongress der Deutschen Gesellschaft für Psychologie in Vienna, 1984. I would like to thank Bernd Behrendt, Wilfried Collet, and Werner Wippich for helpful comments during the 1988 meeting of the Arbeitskreis Angewandte Gedächtnispsychologie in Aachen, FRG.

and goodness-of-fit testing are discussed. Selected applications of uni- and multivariate normal, multinominal, and binomial mixtures to problems of developmental psychology are presented, together with a short overview of computer programs available for mixture distribution analyses.

Many developmental theories share the following features: (a) They are *general* in the sense that they refer to each and every member of an open set of individuals (e.g., the set of all human beings); (b) they are *qualitative* insofar as they presume a set of discrete psychological, biological, or social states; (c) they are *deterministic,* since the possibility of errors (i.e., exceptions from the rule) is not hypothesized by theory formulations; and (d) they contain *theoretical terms* (hypothetical constructs, intervening variables, latent variables, disposition predicates etc.), that is, terms referring to states or events that are not directly observable. Typically these theories claim that all individuals are in either state 1, state 2, . . . , or state c, at any point in time, where the c different states are theoretical, nonempirical categories. Quite often, only some of the set of all possible transitions between states are allowed by theory. For example, often only transitions to "higher" or "better" states are presumed. If no transitions between states are allowed, this is equivalent to the assumption of qualitatively different types of development.

An obvious consequence of the attributes just mentioned is the possibility of decomposing the population of individuals into a finite number of disjunctive and exhaustive classes according to type of development, present-day state, or ontogenetic history of states. Prominent examples of developmental theories following these principles include stage conceptions of cognitive development (cf. Piaget, 1963), theories of moral development (cf. Kohlberg, 1976), theories of information-integration development (cf. Wilkening, Becker and Trabasso, 1980), and theories of learning (cf. Gagné, 1965). These theories use terms like *sensorimotor stage, preconventional level, additive information-integration rule,* or *verbal-associations type of learning,* which are obviously theoretical in the sense specified here. They also maintain that a specific sequence of developmental stages applies to *all* individuals and that inversions of stages *never* occur (cf. Pinard and Laurendeau, 1969).

If we conceive theories of the kind just characterized as fallible, how best to subject them to empirical test? The call for testability implies some kind of "bridge" between the theoretical terms involved and empirical variables observed in experimental or correlational studies. We need *indicators* of the theory's states and/or events. Since "a theory that is not stated in terms of the data analysis to be used cannot be tested" (Guttman, 1981, p. 63), empirical tests are impossible unless explicit indicator variables and relationships between theoretical terms and indicator variables are clearly specified in some kind of formal language.

The traditional approaches to this problem, namely operational definitions and bilateral reduction sentences, have severe drawbacks (see Erdfelder, 1988). Along with the logical problems associated with these methods (Carnap, 1936/1937; Essler and Trapp, 1978), theoretical and empirical categories must be linked by means of *deterministic rules of manifestation,* which demand in a deterministic fashion either unconditionally (in case of operational definitions) or conditionally (in case of bilateral reduction sentences) a one-to-one relation between theoretical and empirical variables. This determinism is equivalent to the assumption of error-free data; *perfect* empirical indicators of theoretical variables are necessary. For developmental psychology at least, this is not a realistic request.

This chapter examines the suitability of a new statistical approach to the problem of theory evaluation in developmental psychology, namely, the analysis of finite mixture distributions (cf. Everitt and Hand, 1981; Titterington, Smith, and Makov, 1985; McLachlan and Basford, 1988). If finite mixture distributions (finite mixtures, for short) are used to test substantive theories, formalization must be done in terms of probability theory. The link between theoretical and empirical variables is established by means of *probabilistic rules of manifestation.* Rules of this kind accept variability in indicator variables even if the values of their theoretical counterparts are fixed. Hence, empirical variables are regarded as imperfect indicators of unobservable, theoretical variables.

This chapter presents an informal characterization (Section 1) and a formal description (Section 2) of finite mixture models. In Section 3 the adequate modeling of developmental theories by finite mixtures is discussed. Section 4 is concerned with the identifiability of finite mixture models and the problems of parameter estimation and goodness-of-fit testing. Applications of finite mixtures to some problems of developmental psychology are considered in Section 5. Section 6 gives a short overview of computer programs available for finite mixture analyses.

1. Testing Deterministic Developmental Theories by Means of Finite Mixture Distribution Analysis

Which features must a statistical model possess to allow adequate formalizations and evaluations of qualitative, deterministic developmental theories? A combination of a *structural model* mimicking the central propositions of the theory with latent variables and a *measurement model* establishing links between latent and empirical variables is necessary. Two conditions of adequacy should be taken into account.

1. The theoretical propositions and the structural model should be *isomorphic*. One aspect of this postulate is that determinism in the theory should lead to determinism in the structural part of the model. Similarly, if the theoretical terms are qualitative, the model latent variables should also be qualitative. Departures from this kind of structural isomorphism involve the risk that a test of the model cannot be interpreted as a test of the theory.

2. The measurement model should be *realistic*. It should not be based on false, presumably false, or even unnecessarily restrictive assumptions concerning the relationships between latent and empirical variables. In general, deterministic rules of manifestation will not fulfill this requirement. A reasonable measurement model should accept measurement error, that is, some amount of variability in empirical indicator variables given fixed values of their theoretical counterparts. Hence, a reasonable measurement model should be based on probabilistic rules of manifestation.

In this chapter it will be shown that if the theory to be evaluated is qualitative and deterministic in nature, a class of statistical models exists that is very well suited to meet all conditions just specified. These models are known as finite mixtures in the statistical literature.

The idea behind this class of models is the following: Individuals (or, more generally, the elements of the empirical sample) may be thought of as members of c different latent classes, each generating a different distribution of observable variables. These variables may be quantitative (continuously distributed) or qualitative (discretely distributed), and with (ordinal level) or without (nominal level) ordered categories. "Latent class" basically means nothing more than "class without perfect empirical indicator(s)." Here we may equate latent classes with groups of individuals within certain psychological, sociological, or biological states, or with a certain history of such states. Fortunately, it is not very important statistically whether the latent classes are conceptualized as categories of one discrete latent variable or as cells of a multivariate latent contingency table. The difference provides only varying parameter restrictions across different classes. These restrictions must be implemented in the latter case and omitted in the former, as will be discussed. Here we presume as many discrete latent variables as there are qualitative dispositional terms in our theory, say l.

Let us consider an example. One hypothesis of Piaget's cognitive developmental theory is that acquired cognitive capabilities never dwindle. Instead, these capabilities are absorbed in the next stage of cognitive development. In the developmental literature this aspect of development is called "integration" (cf. Pinard and Laurendeau, 1969). Specifically, one theorem of Piaget's theory may be stated as follows: "For all human beings x: If x is capable of formal operations, then x is also capable of concrete operations." This proposition comprises $l = 2$ dispositional terms, namely, *capability of formal operations* and *capability of concrete operations*. Thus, the proposition corresponds to a 2×2 latent contingency table comprising one empty cell: The capabilities of concrete and formal operations are the two latent variables and their categories are "present" versus "absent"; the latent class of people capable of formal operations but not capable of concrete operations has to be empty if the proposition is true. Hence, only $c = 3$ latent classes of individuals remain: people possessing both capacities (stage of formal operations), people solely capable of concrete operations (stage of concrete operations), and people capable of neither concrete nor formal operations (stage of sensori-motor intelligence).

If this Piagetian hypothesis holds true, any joint distribution of appropriate indicator variables for the latent capacities (e.g., psychological tests) will reflect the hypothesized subpopulation structure of the total population. In other words, the distribution of observed variables should be representable as a three-component mixture of local distributions within latent classes.

Finite mixture analysis is concerned with the estimation of mixture distribution parameters as well as with the evaluation of the goodness-of-fit of the model.

2. The General Finite Mixture Model

Let us turn now to a more formal treatment of finite probability mixtures. We assume that n different empirical random variables Y_j, $j = 1, \ldots, n$, are given on a common probability space. To simplify notation, we will write the n-dimensional random vector $\mathbf{Y}$. The values of $\mathbf{Y}$ will be denoted by $\mathbf{y}$. We presume that the population consists of c disjunctive and exhaustive latent classes x_k, $k = 1, \ldots, c$. Formally they will be treated as realizations of one discrete latent variable X. By $p(X = x_k)$ we signify the probability that X takes on the value x_k (i.e., the a priori probability of an individual to be a member of the kth latent class).

The probability density function *within* the kth latent class, that is, the function specifying the probability density (or probability, if all n variables are discrete) of $\mathbf{Y} = \mathbf{y}$, given that $X = x_k$, will be denoted by $f(\mathbf{Y} = \mathbf{y} \mid x = x_k)$. It is presumed that $f(\mathbf{Y} = \mathbf{y} \mid X = x_k)$ belongs to a prespecified family of probability density or mass functions and solely depends on some d-dimensional vector of parameters denoted by $\boldsymbol{\theta}_k$, which may vary between the classes. Hence, we may equate

$$f(\mathbf{Y} = \mathbf{y} \mid x = x_k) = g(\mathbf{Y} = \mathbf{y}; \boldsymbol{\theta}_k). \tag{1}$$

For simplicity we shall call $g(\mathbf{Y} = \mathbf{y}; \boldsymbol{\theta}_k)$, as well as the derived functions presented here "density functions," even though some of them are probability mass functions (if $\mathbf{Y}$ is discrete).

Next, using the conditional density $g(\mathbf{Y} = \mathbf{y}; \boldsymbol{\theta}_k)$, we may rewrite the unconditional density of the compound argument ($\mathbf{Y} = \mathbf{y}$ *and* $X = x_k$) as a product

$$f(\mathbf{Y} = \mathbf{y} \text{ and } X = x_k) = g(\mathbf{Y} = \mathbf{y}; \boldsymbol{\theta}_k) p(X = x_k). \tag{2}$$

To get the unconditional density function of the observed variables alone, we simply sum terms across the different latent classes:

$$f(\mathbf{Y} = \mathbf{y}) = \sum_{k=1}^{c} f(\mathbf{Y} = \mathbf{y} \text{ and } X = x_k)$$

$$= \sum_{k=1}^{c} g(\mathbf{Y} = \mathbf{y}; \boldsymbol{\theta}_k) p(X = x_k). \tag{3}$$

This formula represents the general finite mixture distribution model (Everitt and Hand, 1981, p. 4). The random variable $\mathbf{Y}$ is said to have a *finite mixture distribution* and $f(\mathbf{Y} = \mathbf{y})$ is called a *finite mixture density function* (Titterington et al., 1985, p. 1). The names are self-explanatory, insofar as they refer to the fact that the function $f(\mathbf{Y} = \mathbf{y})$ may be viewed as a "mixture" of a finite number of component densities $g(\mathbf{Y} = \mathbf{y}; \boldsymbol{\theta}_k)$, with X as the "mixing variable" and the distribution of X as the "mixing distribution." Sometimes the probabilities $p(X = x_k)$ are referred to as "mixing weights." In the statistical literature, these more neutral terms are preferred to "latent variable," "latent variable distribution," and "latent class probability," respectively.

As may be gathered from equation (3), any finite mixture depends on three classes of parameters: (a) the parameters of the mixing distribution [i.e., the latent class probabilities $p(X = x_k)$]; (b) the elements of the parameter vectors θ_k; and (c) the number c of mixing variable categories. In applications of finite mixtures models, the parameter c, a positive integer value, must be specified a priori, while the parameter classes (a) and (b) often (i.e., if the problem is not ill-conditioned) may be estimated from the data. Of course, this requires a specification of the general model (3). Explicit formulas for the density functions $g(\mathbf{Y} = \mathbf{y}; \theta_k)$ must be inserted, such as uni- and multivariate types of normal, binomial, multinominal, or Poisson density and probability functions. Examples will be considered, along with some general guidelines for appropriate "model building" and a short discussion of some fundamental issues in the analysis of finite mixtures.

3. What Is the Correct Model to Analyze?

Given some qualitative, deterministic hypothesis to test via finite mixture analysis, what is the correct (i.e., hypothesis-isomorph) model to apply to empirical data? First of all, the researcher should clarify what kind of structural model is implied by the hypothesis. A careful analysis of the hypothesis will reveal the "correct" parameter c for the finite mixture model as implied by the hypothesis. Next, the best-known indicators of the latent variables should be selected, and their joint distribution within latent classes should be defined. Last but not least, the possibility of parameter restrictions concerning the latent class probabilities and/or component densities should be realized, if theoretically demanded. We will discuss each of these four steps in more detail.

3.1. Specifying the Parameter c for the Finite Mixture Model

Often, the hypothesis to be tested, if formalized in terms of predicate calculus, will be a special case of the general formula

$$\forall(t)\forall(x)J^l(T_1xt, T_2xt, \ldots, T_lxt), \tag{4}$$

where $\forall$ denotes the universal quantifier, x is a variable referring to individuals, and t is a variable referring to points in time. The $T_k xt$, $k = 1, \ldots, l$, are different two-place dispositional predicates (to read: "individual x has disposition T_k at time t"), and J^l denotes some l-place sentential connective. For any point in time, this proposition corresponds to a latent contingency table of dimension l (one variable for each disposition), comprising 2^l possible cells, since all l variables are dichotomous (representing presence versus absence of a disposition). Typically, the dispositional predicates are linked by J^l in such a way that the resulting proposition is neither contradictory nor tautological. (If this is not the case, the theory may be discarded from the domain of potential valid empirical theories by means of analytical arguments alone.) Hence, at least one cell of the multivariate contingency table must be nonempty, and at least one must be empty, thus reducing the number of possible latent classes to some value c, $1 \leq c < 2^l$. In applications of equation (4), the value c is easily found if the formula is rewritten using the equivalent *disjunctive normal form* (i.e., a disjunction of conjunctions of the l different disposition predicates or of their negations; cf. von Eye and Brandtstädter, 1988). If this transformation is done, the number of elements connected by disjunctions equals the number c of hypothesized latent classes of individuals per point in time. For example: If $l = 2$, and the hypothesis is of the biconditional type "disposition D_1 if and only if disposition D_2," or formally,

$$\forall(x)\forall(t)(D_1 xt \leftrightarrow D_2 xt), \tag{5}$$

then the equivalent disjunctive normal form has $c = 2$ elements connected by disjunction, one describing the presence of both dispositions and one describing the absence of both dispositions:

$$\forall(x)\forall(t)((D_1 xt \text{ and } D_2 xt) \text{ or } (\neg D_1 xt \text{ and } \neg D_2 xt)). \tag{6}$$

It may easily be shown that this procedure is *always* applicable if the hypothesis to be tested is formalizable in terms of first-order predicate logic (cf. Brandtstädter and von Eye, 1982).

A somewhat more complicated situation arises if repeated measurements (longitudinal studies) instead of one-point measurements (cross-sectional studies) are planned or already conducted. While in the latter case the number of latent classes of individuals hypothesized to be nonempty is simply equal to the number of elements in the disjunctive normal form of the hypothesis, in the former case this remains true only under very restrictive assumptions or if each point in time is considered separately. Usually it is desirable to treat the different time points simultaneously. If all kinds of transitions between states across time are allowed, and if the number of theoretically presumed states per time point is s, then, for r repeated measurements (e.g., an r-wave study), s^r different sequences of states are allowed. Hence, an analysis of the joint distribution of all r replications of the empirical variables should be based on a finite mixture model with $c = s^r$ different latent classes.

If the theory imposes restrictions on the possible transitions between states, some value $c < s^r$ will be adequate. This is the case, for example, in stage theories of human development and learning, which prescribe either no transitions or transitions to some "higher" state only. Consider the example of a two-wave study ($r = 2$). If no restrictions are imposed on the possible transitions, $c = s^2$ latent classes must be postulated. On the other hand, if the s states are presumed ordered and only transitions to "higher" states are allowed, this implies a maximum of $c = (s^2 + s)/2$ nonempty latent classes.

Finally, if the theory totally prohibits transitions between states over time (thus defining qualitatively different types of development), the number of latent classes must be equal to the number of states, $c = s$, irrespective of the number of time points.

3.2. Selecting Optimal Indicators of Latent Variables

After specifying the parameter c for the finite mixture model, the researcher should try to select the best-known, available indicator variables for the theory's unobservable dispositions, capacities, states, or events. There is no need to think only of *discrete* empirical variables. Indicators of discrete latent variables may be of any form as long as the theory does not claim otherwise. There is also no need to define only one indicator per latent variable.

Of course, a good indicator should be of high *reliability,* that is, of limited variability when the value of the underlying theoretical variable is fixed. But even more important is the *specificity* of the indicator. Indicators which are sensitive to only one or at most to a few latent variables should be chosen over indicators covarying with a huge number of latent variables. Specificity allows certain kinds of a priori restrictions on the parameter vector $\boldsymbol{\theta}_k$, thus strengthening the testability of the model.

3.3. Specifying a Reasonable Local Distribution Model

After selecting the indicator variables, the researcher should choose a reasonable model for the *local* distribution of indicator variables, that is, for the distribution within latent classes. Some kind of discrete distribution model should be chosen if the indicator variables may assume only a finite number of discrete values. This will be true if the variables are dichotomous (e.g., of "yes"/"no" type), polytomous with unordered categories (e.g., Catholic, Jewish, etc.), or polytomous with ordered categories (e.g., very, somewhat, little). On the other hand, if there is an infinite number of possible realizations of the empirical variables, then a continuous local distribution model should be chosen. Furthermore, the researcher must determine whether the chosen distribution model is incorrect on a priori grounds. It is nonsense, for instance, to select a normal distribution model for continuous variables bounded in the closed real interval $[0, 1]$ (e.g., relative frequencies) or for variables which may take on values solely in the positive reals (e.g., response times). For these kinds of variables, some kind of transformation (e.g., logistic or logarithmic) might help, such that the normal distribution model, which would be wrong for the original variables, would perhaps be appropriate for the transformed variables.

Note that when there is more than one empirical variable, *multivariate* densities $g(\mathbf{Y} = \mathbf{y}; \boldsymbol{\theta}_k)$ must be specified. Knowledge of reasonable *univariate* local distribution models for each empirical variable considered in isolation does not suffice to define an appropriate *multivariate* local distribution model. Often the principle of *local stochastic independence* will help. If the empirical variables may be presumed locally (i.e., within latent classes) stochastic independent, then the joint multivariate density $g(\mathbf{Y} = \mathbf{y}; \boldsymbol{\theta}_k)$ is obtained from the univariate densities $g(Y_j = y_j; \boldsymbol{\theta}_k)$ simply by multiplication:

$$g(\mathbf{Y} = \mathbf{y}; \boldsymbol{\theta}_k) = \prod_{j=1}^{n} g(Y_j = y_j; \boldsymbol{\theta}_k). \tag{7}$$

The idea underlying the assumption of local stochastic independence is that all dependencies between observed variables may be explained by the latent class structure of the population. Dependencies among empirical variables may therefore be regarded as some kind of "artifact" produced by aggregating individuals across different latent classes.

The assumption of local stochastic independence implies that the variables in question do not measure unknown attributes on which individuals within latent classes differ. If there are such attributes (and very likely there will be), and if at least two of the observed variables are sensitive to those attributes, then local dependencies between these variables result. Hence, since complete uniformity within latent classes cannot reasonably be assumed, the researcher who makes use of the assumption of local stochastic independence must try to select indicators which are (a) very sensitive to the latent variables they are presumed to measure and are at the same time (b) very insensitive to other attributes on which individuals might differ. In the terminology of Campbell and Fiske (1959), the indicators should have high convergent as well as high discriminant validity.

There are alternatives to the assumption of local stochastic independence. Sometimes the best imaginable latent class indicator is not a special marginal distribution of *one* empirical variable. Instead, special kinds of dependencies between *two or more* empirical variables may be more reasonably modeled. It is possible, for example, that a theory prescribes a positive correlation between two observed variables Y_1 and Y_2 within latent class 1, a zero correlation between the variables within latent class 2, and a negative correlation within latent class 3. An appropriate label for latent variables moderating the type of dependency between manifest variables is *latent moderator variable*. As will become clear, models with latent moderator variables are easily analyzed within the framework of finite mixtures if the latent moderator variable is assumed to be discrete.

To summarize, the assumption of local stochastic independence for *univariate* local distributions is only one among many possibilities that may be used to derive a *multivariate* local distribution model. In general, given certain kinds of marginal distributions, only a few additional restrictions concerning the joint distribution (e.g., only linear dependencies are allowed) will suffice to obtain an analyzable model.

3.4. Imposing Parameter Restrictions

Finally, the last preparatory step is to choose appropriate parameter restrictions. Restrictions concerning the latent class probabilities $p(X = x_k)$ are possible, but are seldom used in practice. If some theory demands, for example, that in a given population latent class 2 is twice as probable as latent class 1, then the restriction $p(X = x_2) = 2p(X = x_1)$ should be implemented in the model. More frequent are theoretically demanded restrictions concerning the parameters of the component densities (i.e., the parameter vectors θ_k, $k = 1, \ldots, c$). Consider, for example, a hypothesis made of l theoretical terms. Suppose that one indicator variable per dichotomous latent variable has been measured ($n = l$) and that each Y_j indicates one and only one latent variable. Then the following must be true: The local distribution of each Y_j varies between two arbitrarily selected latent classes x_{k1} and x_{k2}, if and only if x_{k1} and x_{k2} correspond to different values on the latent variable Y_j indicates. Thus, the parameters of the local distribution *may not* vary between classes belonging to the same level of the latent variable Y_j indicates. This implies a set of equality constraints on the elements of the parameter vectors θ_k which will be different for different indicators.

In general, equality constraints will always be necessary if different latent variables are measured by different, a priori specified indicators. Imposing classes of equality constraints is the correct way to define which observable variable(s) measure(s) which latent variable(s). The rule of thumb is easy. To define a group of empirical variables as indicators of a particular latent variable, restrict the parameters of their local distributions as equal between latent classes taking on the same latent variable value. If equality constraints are omitted, an ambiguous measurement model will be the consequence: Each observable variable may indicate an arbitrary number of latent variables. Nevertheless, such models may be testable and identifiable; there therefore is no reason to discard them on a priori grounds.

4. Statistical Problems in Finite Mixture Analyses

In analytical and empirical analyses of finite mixture models the researcher should follow three steps.

1. Check the identifiability of the model. Does a *unique* set of parameters c, θ_k, and $p(X = x_k)$ define each member of the class of mixtures considered?

2. If identified, estimate the parameters θ_k and $p(X = x_k)$. If all parameters are unrestricted, dc conditional distribution parameters θ_k must be estimated. Since the latent class probabilities $p(X = x_k)$ must sum to unity, one of the c probabilities is redundant; hence, only $c - 1$ latent class probability parameters are free to vary. In total, $c(d + 1) - 1$ parameters must be estimated if no constraints are imposed on the model.

3. Test the goodness-of-fit of the model. Prior to the interpretation of the parameter estimates from a substantive perspective, the researcher must check to see that the empirical data structure "obeys" the specified mixture model. Only when data and model correspond closely does the interpretation of parameter estimates make sense. While parameter estimation without some kind of goodness-of-fit check is useless in

general, the reverse does not hold. Sometimes all we need to know is if the mixture model is empirically adequate, irrespective of the values of the parameters. Thus, from a theory-testing point of view, the goodness-of-fit test is the most important of the three steps mentioned.

4.1. The Problem of Indentifiability

Unlike the estimation of parameters and the test of goodness-of-fit, the check of identifiability concerns only the model. This is the only step of finite mixture analyses which may be performed without reference to a special set of data. To illustrate the problem, consider a simple example (taken from Everitt and Hand, 1981, p. 5): Let $U(Y = y; a, b)$ denote the uniform density function in the real interval $[a, b]$, such that

$$U(Y = y; a, b) = \begin{cases} \dfrac{1}{b-a}, & \text{if } a \leq y \leq b, \\ 0, & \text{else.} \end{cases} \qquad (8)$$

It is easy to show that, for instance, the two-component uniform mixture

$$f_1(Y = y) = \tfrac{1}{3}U(Y = y; -1, 1) + \tfrac{2}{3}U(Y = y; -2, 2)$$

cannot be distinguished from the mixture

$$f_2(Y = y) = \tfrac{1}{2}U(Y = y; -2, 1) + \tfrac{1}{2}U(Y = y; -1, 2),$$

that is, $f_1(Y = y) = f_2(Y = y)$. The graph corresponding to this density function is illustrated in Figure 14.1.

If, as in this example, two or more different constellations of parameters c, $p(X = x_k)$, and θ_k, $k = 1, \dots, c$, generate the same mixture distribution, we say that the mixture is *not identifiable*. More formally we may define (see Everitt and Hand, 1981, p. 5):

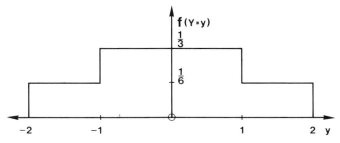

FIGURE 14.1. Graph of an unidentifiable mixture of uniform density functions.

Definition. A class D of finite mixture density functions is said to be *identifiable*, if and only if for all $f_i(\mathbf{Y} = \mathbf{y})$ and $f_{i'}(\mathbf{Y} = \mathbf{y})$ in D the equality

$$f_i(\mathbf{Y} = \mathbf{y}) = \sum_{k=1}^{c} g(\mathbf{Y} = \mathbf{y}; \boldsymbol{\theta}_k) p(X = x_k)$$

$$= f_{i'}(\mathbf{Y} = \mathbf{y}) = \sum_{k'=1}^{c'} g(\mathbf{Y} = \mathbf{y}; \boldsymbol{\theta}'_{k'}) p'(X = x_{k'})$$

implies

1. $c = c'$.
2. For all k there exists some k' such that

$$p(X = x_k) = p'(X = x_{k'}) \qquad \text{and} \qquad \boldsymbol{\theta}_k = \boldsymbol{\theta}'_{k'}.$$

Basically this definition states that parameters of identifiable mixtures are unique up to permutations of the latent class indices. Of course, an equivalent definition may be given in terms of finite mixture cumulative distribution functions (c.d.f.'s) derivable from the density functions by integration (see Titterington et al., 1985, p. 36).

Important theoretical results concerning the identifiability of mixtures have been presented by Teicher (1961, 1963) and Yakowitz and Spragins (1968). The following theorem is due to the last authors:

Theorem. *The class of all finite mixtures of a set of component densities $g(\mathbf{Y} = \mathbf{y}; \boldsymbol{\theta}_k)$ is identifiable if and only if the set of c.d.f.'s $G(\mathbf{Y} = \mathbf{y}; \boldsymbol{\theta}_k)$ corresponding to the component densities is linearly independent; that is, if and only if the linear equation*

$$\sum_{k=1}^{c} a_k G(\mathbf{y} = \mathbf{y}; \boldsymbol{\theta}_k) = 0, \qquad a_k \in \mathbb{R},$$

has no other than the trivial solution $a_1 = a_2 = \cdots = a_c = 0$.

The proof of this theorem is given in Everitt and Hand (1981, pp. 6–7) and Titterington et al. (1985, p. 37). Using this (and related) theorems, one can demonstrate the identifiability of important classes of finite mixtures, such as mixtures of uni- and multivariate normal distributions, gamma distribution mixtures, exponential mixtures, Poisson mixtures, Cauchy mixtures, and other kinds of mixtures (cf. Teicher, 1961, 1963; Yakowitz and Spragins, 1968; Titterington et al., 1985). Furthermore, Teicher (1967) has shown that every multivariate mixture derived from univariate mixtures under the assumption of local stochastic independence is identifiable if and only if the corresponding univariate mixtures are identifiable.

As can be seen, much of the necessary analytical work concerning identifiability of finite mixtures has already been done. The researcher who wants to make use of one of the mixture models mentioned above need not bother about identifiability. On the other hand, mixtures of uniform distributions as well as mixtures based on discrete distributions, for instance binomial mixtures and mixtures of products of multinominal distributions (i.e., latent class models), are examples of models that are not identifiable in general. Despite this, it is possible to define subclasses of binomial mixtures and of latent class models that are identifiable. A *necessary condition* for such a class of mixtures to be identifiable is that the number of parameters of the mixture model may not exceed $m - 1$, where m denotes the number of categories of the resulting mixture distribution. (The number of parameters has to be one less than m, since only $m - 1$ of the probabilities associated with the m categories of the discrete mixture are free to vary.) This suggests at least two ways of transforming not identifiable models into identifiable models: (a) Increasing the number of empirical indicators or the number of categories per indicator, such that finally there are more categories than parameters; (b) decreasing the number of parameters by imposing some kind of parameter restriction. Note however that the above mentioned condition is in general not sufficient, that is, there is no guarantee that a model with less than $m - 1$ parameters is identifiable.

An important question remains to be answered. Are not identifiable models generally useless? It is well known that not identifiable models may nevertheless be testable (cf. Bamber and Van Santen, 1985). Consider the example of uniform mixtures discussed before. It is possible that some kind of substantive hypothesis implies exactly the density illustrated in Figure 14.1. The density corresponds to a mixture model which is not identifiable, as we know now, but this model is nevertheless testable: Empirical data may obey the postulated density or they may deviate from it significantly. Hence, if we want to know whether the hypothesis under consideration is true or not, we must perform some kind of goodness-of-fit check based on a given empirical histogram. This example shows that not identifiable models may be very important. It is sometimes possible and necessary to test them. The fact that a model, as derived from substantive theory, is not identifiable, does not disqualify the hypothesis or theory from which the model was derived. It simply implies that in this case parameter estimation as well as parameter interpretation does not make sense. If this rule is observed, it is recommended that models (if possible) as derived from a given theory be tested even if they are not identifiable. From a theory-testing point of view it is much worse to force identifiability by some kind of unjustified parameter restriction.

4.2. The Problem of Parameter Estimation

In this section, the problem of parameter estimation, based on samples randomly drawn from the mixture population, is discussed. We consider only the estimation of (constrained or unconstrained) latent class probabilities as well as of (constrained or unconstrained) parameters of the component densities. It is presumed that the number of latent classes (i.e., the parameter c of the model) is fixed to some definite value, according to a theory or hypothesis. Whether this parameter fixation is empirically adequate is the subject of the next section.

Among the different methods of parameter estimation used in connection with finite mixtures, the *method of maximum likelihood* (*ML method*) clearly has the most advantages (cf. Erdfelder, 1988; Everitt & Hand, 1981; Titterington et al., 1985). ML estimation is based on the assumption of independent sampling from the mixture distribution, such that the probability density of the total sample equals the product of the densities of single elements from the sample. The method involves the maximization of the likelihood function

$$L := \prod_{i=1}^{N} f(\mathbf{Y} = \mathbf{y}_i)$$

$$= \prod_{i=1}^{N} \sum_{k=1}^{c} g(\mathbf{Y} = \mathbf{y}_i; \boldsymbol{\theta}_k) p(X = x_k), \tag{9}$$

where $\mathbf{y}_i$ denotes the realization of $\mathbf{Y}$ for the ith element of the sample, and N denotes the total sample size. Hence, values for the parameters $p(X = x_k)$ and $\boldsymbol{\theta}_k$ must be selected (i.e., *ML estimators*) such that L has a global maximum, given a particular set of data.

There is no canonical way to find the global maximum of L. First, differentiating L successively with respect to the parameters and equating all partial derivatives to zero yields an equation system which has no explicit solution. Second, even if we somehow arrive at a solution, the solution does not necessarily define a set of ML estimators, since it may correspond to a minimum, saddle point, or local (but not global) maximum of the likelihood function. Fortunately, an intuitively appealing algorithm exists, which circumvents most of these problems. Hasselblad (1966, 1969) was the first to propose this numerical method, which later became known as the estimation maximization algorithm (EM algorithm). Dempster, Laird and Rubin (1977) published important theoretical results concerning the behavior of the EM algorithm in the wider context of maximum likelihood estimation from incomplete data. In particular, the authors proved the following theorems:

1. The EM algorithm is monotone, that is, in each step of the algorithm the likelihood of the (preliminary) parameter estimates increases.

2. Under very mild regularity conditions the EM algorithm converges to some point within the parameter space.

3. In almost all applications, the convergence point corresponds to a *local, if not global*, maximum of the likelihood function.

The EM algorithm proceeds as follows: In the first step, given a set of initial estimates (e.g., moment estimators) for the parameters, the complete data set is estimated (E-step). Estimation of the complete data set involves estimating the probability that some element i ($i = 1, \ldots, N$) from the sample belongs to latent class x_k ($k = 1, \ldots, c$) given a certain realization $\mathbf{Y} = \mathbf{y}_i$ of the observed variables and given certain preliminary estimates of parameters. If we denote the estimates in cycle l of the EM algorithm by $\hat{p}(X = x_k)^{(l)}$ and $\hat{\boldsymbol{\theta}}_k^{(l)}$ respectively, then, using Bayes' theorem, the desired conditional probabilities may be computed as follows:

$$\hat{p}(X = x_k \mid \mathbf{Y} = \mathbf{y}_i; \hat{p}(X = x_k)^{(l)}, \hat{\boldsymbol{\theta}}_k^{(l)})$$

$$= \frac{\hat{p}(X = x_k)^{(l)} g(\mathbf{Y} = \mathbf{y}_i; \hat{\boldsymbol{\theta}}_k^{(l)})}{\sum_{k=1}^{c} \hat{p}(X = x_k)^{(l)} g(\mathbf{Y} = \mathbf{y}_i; \hat{\boldsymbol{\theta}}_k^{(l)})}. \tag{10}$$

In the second step of the algorithm (M-step), the estimated complete data set is used to estimate new parameter values $\hat{p}(X = x_k)^{(l+1)}$ and $\hat{\boldsymbol{\theta}}_k^{(l+1)}$, which will increase the likelihood. This is done in a manner similar to the usual ML estimation for complete data sets, except that each observation $\mathbf{y}_i$ contributes to the parameter estimates for different latent classes with different weights, each proportional to the probability estimates $\hat{p}(X = x_k \mid \mathbf{Y} = \mathbf{y}_i; \hat{p}(X = x_k)^{(l)}, \hat{\boldsymbol{\theta}}_k^{(l)})$. Parameter restrictions may be implemented as easily in the M-step of the algorithm as in the usual ML procedures for complete data sets. Further details of the parameter estimation procedure are described by Hasselblad (1969), Wolfe (1970), and Dempster et al. (1977).

The E-step and the M-step of the EM algorithm are repeated iteratively until convergence to some stationary point of the likelihood function is attained. Unfortunately, the speed of the EM algorithm is not very satisfactory in general. Many iteration cycles may be necessary until convergence is attained (cf. Titterington et al., 1985, pp. 89–91). While there do exist faster methods of ML estimation (e.g., variants and combinations of Newton–Raphson and the method of scoring), these methods are by far more complicated in the numerical procedures involved. Furthermore, there is not guarantee of monotone behavior (see Titterington et al., 1985). These are strong arguments to rely on the EM algorithm, at least when very fast computers are available, such that slow convergence need not bother the user.

Unfortunately, there are other problems inherent to the ML method which cannot be ruled out. Small sample sizes often induce "ugly" likelihood surfaces containing multiple local maxima. Then the EM algorithm, like all other hill-climbing procedures, may run into a local, but not global, maximum. One way to reduce the probability of falsely accepting a local as a global maximum is to start the EM algorithm several times using different starting values. There is no analytical way to decide which method generally generates the best initial parameter estimates, such that the starting values are already close to the correct ML solution. However, if computation time is likely not to be a problem, it is surely good advice to use several of the methods proposed in the literature in combination [e.g., moment estimators; minimum distance estimators; decomposition of the sample by some kind of cluster analysis, (e.g., Titterington et al. 1987; Wolfe, 1978)].

One special problem of ML estimation in finite mixture analyses was not mentioned until now: In some situations the likelihood function L is not bounded, so that parameter constellations exist that are associated with a likelihood value of infinity. For example, if we consider a c-component univariate normal mixture, the likelihood L approaches infinity whenever the standard deviation of one of its components approaches zero. Hence, the ML estimate does not exist. But does the ML method break down in such cases, as some authors (e.g., Day, 1969, and Johnson and Kotz, 1972) have concluded? Clearly solutions suggesting zero variances within latent classes are pathological (Everitt and Hand, 1981, p. 9). In such cases we are forced to discard the global maximum. The alternative is to look for the highest local maximum in the infield of the parameter space. Although it does not define a set of ML estimators, it seems to define the best nondegenerative solution (cf. Duda and Hart, 1973). Note also that the problem may be ruled out completely if we introduce hypotheses that impose certain restrictions on the standard deviations within latent classes, such that they cannot take on zero values (cf. Hathaway, 1985). Thus, this problem does not force us to use alternative methods of estimation proposed in the literature (cf. Titterington et al., 1985).

4.3. The Problem of Goodness-of-Fit Testing

What possibilities exist for testing the statistical null hypothesis that a given sample was drawn from a theoretically derived mixture population? Provided that the sample size is large enough to justify the application of asymptotic theory results, one may think of the likelihood ratio chi-square approach. Let L_0 and L_1 denote the sample-based likelihood values for the null hypothesis model and some alternative model, respectively, such that the former is a special case of the latter. The likelihood ratio (LR) statistic may then be written as follows:

$$\text{LR-}\chi^2 := -2 \ln\left(\frac{L_0}{L_1}\right). \tag{11}$$

Given the validity of the null hypothesis model, this statistic is asymptotically distributed as chi-square, provided that some regularity conditions are met, first described by Wilks (1938). The degrees of freedom (df) of the asymptotic distribution correspond to the number of *independent* constraints necessary to transform the more general model into the null hypothesis model, that is, the number of free parameters in the alternative model minus the number of free parameters in the null hypothesis model.

Since the likelihood values necessary to compute the statistic are by-products of the ML estimation procedure, the statistic is computationally easy to apply. At least in situations where reasonable alternative models can be defined, the LR-χ^2 statistic seems to offer a simple and routine way of testing finite mixture models as derived from a substantive theory. For example, if some theory postulates a two-component normal mixture with equal means in each of its components, we could select some unrestricted two-component normal mixture as an alternative model (i.e., df = 1). It would especially be wise to do so if the information about means is highly critical, whereas the value of c is not.

This could be the case, for example, if previous empirical research or analytical arguments suggest the validity of the parameter $c = 2$ but leave open the question as to whether or not the two means are equal.

What should we do, however, if the value of c is critical? As already shown, in cases where finite mixtures are used to test deterministic developmental theories this will be the standard situation. One might be tempted to apply the LR-χ^2 statistic here too. If an unrestricted two-component normal mixture is the theoretically derived null hypothesis model, we could select some higher-component mixture, say, a three-component normal mixture, as the more general alternative model. One difficulty with this approach becomes obvious at once: There is no unique way of transforming the more general model into the null hypothesis model, so that our rule to compute the degrees of freedom becomes dubious. We could introduce the restriction $p(X = x_3) = 0$ into the alternative model to get the null hypothesis model; this suggests df $= 1$. But the same result could be achieved by equating the means and the standard deviations of, say, the second and third components, which suggests df $= 2$ (cf. Titterington et al., 1985, pp. 4–5). This problem is a consequence of the fact that one latent class probability lies at the boundary of the parameter space of the more general model if the null hypothesis model is valid. Obviously this will always be true if we compare mixtures with different numbers of components.

Unfortunately the regularity conditions necessary to derive the asymptotic chi-square distribution of the LR-χ^2 statistic are not fulfilled in this situation (cf. Aitkin, Anderson and Hinde, 1981; Aitkin and Wilson, 1980; Binder, 1978; Titterington et al., 1985, pp. 152–156). Hence, it is not at all clear how to evaluate an empirical LR-χ^2 result. The same problem may occur in other model frameworks too, usually without being noticed. For example, Andres and Haessner (1984) showed that some simple LISREL models generate LR-χ^2 statistics that are not asymptotically chi-square distributed. Moreover, the same authors proved that the correct asymptotic distribution is sometimes stochastically larger, and sometimes stochastically smaller than the chi-square distribution assumed by the LISREL program. The computed p-value is therefore in general neither an upper bound nor a lower bound of the correct p-value.

If the asymptotic distribution is not chi-square, and chi-square-based upper bounds on the p-values cannot be derived, it is desirable to find some transformation T of LR-χ^2 such that $T(\text{LR-}\chi^2)$ is (approximately) chi-square distributed. Wolfe (1978) proceeded along these lines. For the special case of testing a c-component d-variate normal mixture (null hypothesis) against a c'-component d-variate normal mixture ($c' > c$), he proposed the modified statistic

$$\frac{N - 1 - d - 0.5c'}{N} \text{ LR-}\chi^2, \tag{12}$$

which, according to a set of simulation results, is approximately chi-square distributed with df $= 2d(c' - c)$ if the null hypothesis is true. Another Monte Carlo study (Everitt, 1981) showed however that for $c = 1$ and $c' = 2$ either $N > 50$ or $N > 10d$ is necessary for acceptable behavior of the proposed approximation. Moreover, even if these conditions are met, the power of the test is fairly low unless the two components' means are very well separated.

We thus have to admit that there is no generally justified way of testing goodness-of-fit based on the LR-χ^2 statistic. Some authors (e.g., Everitt and Hand, 1981, p. 118) nevertheless recommend the routine application of the likelihood ratio test with chi-square as the reference distribution, perhaps because they feel that the error made by referring to chi-square is not very great. A safer procedure for finite mixture goodness-of-fit testing is probably the following:

1. First check, if a statistically justified formal goodness-of-fit test exists for your special test problem. In some cases this may be the standard likelihood ratio test (for instance, if models with the same parameter c, but different restrictions on the within class distributions, are tested against each other). Even if models with different parameters c and c' are to be tested against each other, there exist some statistically "clean" tests for selected cases (see Titterington et al., 1985, pp. 149–159, for details). Apply one of these tests whenever possible.

2. If a statistically clean test does not exist, apply the standard likelihood ratio test *in combination with some informal evaluation techniques.* Among these are graphical checks of the correspondence between empirical and fitted histograms, so-called probability plots (cf. Everitt and Hand, 1981, pp. 108–116), informal searches for discontinuities in plots of the log-likelihood against the number c of assumed latent classes, and other sophisticated graphical methods (cf. Titterington et al., 1985, chaps. 4.1, 4.7).

3. In very important applications perhaps rely upon the Monte Carlo method (cf. McLachlan and Basford, 1988). For example, Aitkin et al. (1981) first estimated the parameters of a special latent class model and

then simulated random samples of the same size as the original data set, inserting the empirical parameter estimates as (population) parameters. If for every simulated sample a LR-χ^2 statistic is computed, the sampling distribution of χ^2 results might be used as a reference distribution for the empirical LR-χ^2 statistic value. Of course, this procedure demands a great deal of computational effort and therefore cannot be recommended for routine applications. Moreover, although the procedure has some plausibility, it is not clear if the true asymptotic distribution of the LR-χ^2 statistic is invariant under changes in the values of free parameters of the null hypothesis model. This is obviously a premise of the procedure selected by Aitkin et al. (1981).

We must conclude that the state of the art in goodness-of-fit testing of finite mixture models is far from perfect. It is the experience of the author, however, that clever combinations of formal and graphical evaluation techniques will yield a relatively clear picture of empirical model adequacy/inadequacy in nearly all applications.

5.   Special Applications of Finite Mixtures

5.1.   Mixtures of Univariate Normal Distributions

Isaac and O'Connor (1969) were the first to apply finite mixture models in developmental psychology. Their goal was to test a discontinuity theory of psychological development, based on the notion of six "levels of abstraction."

The data used by the authors were responses to several problem solving tasks. First the authors presented three "level-4 tasks" of medium difficulty to samples of 300 to 500 university students. The histograms resulting from an ad hoc scoring of subjects' responses were separately fitted by four-component normal mixture models using Hasselblads (1969) EM method. The fit obtained was in each case considered satisfactory by the authors. A somewhat simpler problem solving task designed especially for the third level of abstraction yielded an acceptable fit for a three-component normal mixture model with 12- to 17-year-old pupils as subjects ($N = 264$). Finally two "level-2 experiments", each conducted with about 200 children 6 to 10 years old, resulted in good-fit indices for two-component normal mixture models. The same problems presented to five-year-old subjects could not be analyzed since many children did not succeed in solving the problems. Unfortunately, very complicated problems allowing the test of five- or even six-component hypotheses were not investigated by the authors. Therefore, the hypothesis of exactly *six* stages of development cannot be regarded as empirically supported by their work. Nevertheless, the notion of qualitatively different levels of abstraction, developing step by step in the ontogenesis of men, and measurable by different kinds of problem-solving tasks, may be regarded as empirically fruitful and partially supported by means of finite mixture analyses.

One special problem in the work of Isaac and O'Connor (1969) is their confusion concerning the relation between the multimodality of distributions and the presence of a mixture distribution. The authors designate their hypothesis as a "multimodality hypothesis," and they conclude that multimodality has been confirmed because multiple normal mixtures yield acceptable fits. However, as Murphy (1964), among others, has shown, neither does the presence of a mixture imply multimodality, nor does multimodality (especially in a sample histogram) imply the presence of a mixture. Some beta-distributions, for instance are bimodal without being finite mixtures. On the other hand, two-component normal mixtures must be unimodal whenever the smaller of the two standard deviations is greater than half of the absolute mean difference between the components (Behboodian, 1970). Hence, the two concepts are unrelated in general. The researcher must decide explicitly if he or she is interested in the number of modes or the number of components of a distribution. If the task is to test discrete deterministic hypotheses, such as Isaac and O'Connor's (1969) hypothesis, the number of modes will usually be of minor importance.

As one might expect, finite mixture models are also very well suited to test certain genetic theories of development. Thomas (1987), for example, presented a quite simple but nevertheless empirically very successful genetic theory of high aptitude development using a normal mixture model. A summary is given by Erdfelder (1988).

5.2. Mixtures of Multivariate Normal Distributions

If more than one quantitative empirical indicator variable for the theoretically presumed latent types or states is available, multivariate normal mixture models often arise naturally from univariate mixtures. In the work of Isaac and O'Connor (1969), for example, several tasks for the second and for the fourth level of abstraction were presented to the subjects. However, *different* samples of subjects worked on the (presumably) parallel task. In the case of a repeated-measurement design for these tasks, the authors alternatively could have selected multivariate normal mixture models to analyze the data for all parallel tasks simultaneously.

This possibility raises an important question: How should one proceed when univariate as well as a multivariate mixture analyses are in accordance with the theory under investigation? Well, multivariate modeling should be preferred because it guarantees a *stronger* test of the theory. It is well known, for example, that under certain conditions (e.g., all variables standardized) the univariate marginal distributions of bivariate normal mixtures can be equal to simple (one-component) normal distributions (cf. Johnson and Kotz, 1972, p. 74; Titterington et al., 1985, p. 68). Thus if a bivariate normal mixture, say, with $c = 2$ latent classes underlies the data, if a univariate approach to the data is selected, a test of a model presuming $c = 1$ latent classes could systematically lead to an acceptance of the wrong model.

Finite mixture models with homogeneous (nonmixture) margins are quite interesting from another point of view. Consider the special case of a two-component bivariate mixture of standardized normals with correlation ρ_1 in the first component and correlation $\rho_2 = -\rho_1$ in the second component. For correlations significantly different from zero, such a model will generate bivariate data sets that look like fuzzy X's in the bivariate scattergram. If the correlations tend to 1 and -1, respectively, the scattergram becomes a perfect X, centered at the origin of the coordinate system (cf. Figure 4.1.10 in Titterington et al., 1985, p. 68). Data structures like these immediately suggest the idea of two qualitatively different subsamples: one with a positive, and one with a negative linear regression among the observed variables Y_1 and Y_2. To put it another way, there seems to be a dichotomous latent moderator variable underlying the data structure, the values of which determine the type of regression from Y_1 on Y_2. Thus, at least in this context, finite mixture analysis may mean latent moderator analysis.

5.3. Mixtures of Products of Multinominal Distributions: The Latent Class Model

The most important mixture model for developmental psychologists, as reflected by the number of published empirical applications, seems to be the latent class model. Since reviews for developmental psychologists already exist (e.g., Bergan, 1988; Rindskopf, 1987, this volume), this

model will not be discussed here in detail. A short integration of latent
class analysis (LCA) in the framework of finite mixture analysis will
suffice.

Readers familiar with the notation commonly used in connection with
LCA but unfamiliar with the notation used in this chapter will perhaps
have some difficulties in realizing that the latent class model is a special
case of the more general finite mixture model as introduced in equation
(3). Using the notation used by Rindskopf in this volume, we write

$$p(X = x_t) = \pi_t^X, \qquad t = 1, \ldots, T \tag{13}$$

for $c = T$ latent class probabilities and

$$g(\mathbf{Y} = \mathbf{y}; \boldsymbol{\theta}_t) = \pi_{it}^{AX} \pi_{jt}^{BX} \pi_{kt}^{CX} \pi_{lt}^{DX} \tag{14}$$

for the conditional densities within latent classes. Here, $\mathbf{Y}$ is a vector of
four discrete random variables A, B, C, and D comprising I, J, K, and L
categories, respectively. Insertion of equations (13) and (14) in equation
(3) transforms the general finite mixture model into a latent class model
for four-dimensional contingency tables. Generalization to latent class
models for tables of higher and lower dimensionality is obvious and need
not be discussed here.

What are the core assumptions in latent class models? First of all, the
empirical indicator variables are presumed to be *qualitative,* in contrast
to the quantitative empirical variables we have dealt with in the last two
sections. The unrestricted latent class model imposes no constraints
whatsoever on the local marginal distributions of the empirical variables
within latent classes. They are assumed to be multinomial and hence are
characterized by the category probabilities π_{it}^{AX} ($i = 1, \ldots, I - 1$), π_{jt}^{BX}
($j = 1, \ldots, J - 1$), π_{kt}^{CX} ($k = 1, \ldots, K - 1$), and π_{lt}^{DX} ($l = 1, \ldots, L - 1$).
(In each case probabilities for the last category need not be specified,
since the probabilities must sum to unity.) Note that the local distributions are *fully* characterized by these probabilities [see equation (14)].
Hence, the second core assumption in latent class models is that the joint
distribution of the manifest variables within latent classes is completely
determined once the marginal distributions are specified. Some people
prefer to say that the *assumption of local stochastic independence* is at the
heart of LCA. At first sight this seems to be the better formulation
because it gives an explicit justification for the product rule used to
calculate the joint distribution from the marginals. However, this
formulation is slightly misleading because it suggests that local stochastic

TABLE 14.1. Hypothetical Joint Distribution of Two Dichotomous Variables A and B within the tth Latent Class

	B		
A	1	2	Sum
1	$p_t q_t$	$p_t(1-q_t)$	p_t
2	$(1-p_t)(1-q_t)$	$(1-p_t)q_t$	$1-p_t$
Sum	$p_t q_t + (1-p_t)(1-q_t)$	$p_t(1-q_t) + (1-p_t)q_t$	1

independence is a *conditio sine qua non* of LCA applications. This is wrong, as we shall see in a moment.

First, suppose that violations of local stochastic independence in the $ABCD$ table are due to bivariate dependencies between, say, C and D exclusively. We could then easily generate a new variable E, the categories of which are composed of all possible combinations of the categories of C and D. By analyzing the ABE table instead of the original $ABCD$ table, we circumvent the independence assumption (cf. Formann, 1984, chap. 6.4).

As another example, consider a model for two dichotomous variables A and B with two latent classes. Let the local joint distribution within the tth latent class ($t = 1, 2$) be of the multinominal type specified in Table 14.1. It can be verified that for $p_1 = p_2 = 0.5$ and $q_1 = (1 - q_2) = 0.9$, to give just one example, this model is somewhat analogous to the latent moderator model mentioned earlier. Whereas the correlation between A and B is positive in the first latent class, it is negative in the second. Since A and B are obviously *not* locally stochastic independent one might falsely conclude the the model cannot be analyzed by means of LCA. However, if we define a new variable B^* by

$$B^* := \begin{cases} B, & \text{iff } A = 1, \\ 3 - B, & \text{otherwise,} \end{cases} \qquad (15)$$

and look at the AB^* table, we see that the model has been transformed into a latent class model with local stochastic independence. This is illustrated in Table 14.2.

It need not bother us here that without further restrictions the model is not identified. It is possible to force identifiability by fixing parameter

TABLE 14.2. Joint Distribution of A and B within Latent Class t after Switching of the Categories of B within the Second Category of A

	B^*		
A	1	2	Sum
1	$p_t q_t$	$p_t(1-q_t)$	p_t
2	$(1-p_t)q_t$	$(1-p_t)(1-q_t)$	$1-p_t$
Sum	q_t	$1-q_t$	1

The recoded variable is denoted by B^*.

values. The important fact is that some models violating local stochastic independence can be analyzed by means of LCA if one makes use of certain transformations of manifest variables.

5.4. Mixtures of Binomial Distributions

Sometimes qualitative empirical variables have a natural order imposed on the categories. Rating scales, for example, often reflect the degree of agreement with a certain statement, such that a larger number means more agreement than a lower number. Counts are another example. The number of incidences of a certain event is a qualitative ordered variable, since higher numbers reflect higher incidence rates. An unconstrained latent class model is obviously not appropriate here; order information is lost when multinominal marginal distributions within latent classes are assumed. A normal mixture model, on the other hand, is also inappropriate because rating scales, counts, and other qualitative ordered variables have only a finite number of discrete realizations. Ordered variables with a low number of distinct categories are frequently used in developmental research, and a fairly satisfactory local distribution model must be found for them.

For the univariate case, the local binomial distribution is often appropriate. Let the qualitative ordered variable Y have $m + 1$ categories $Y = 0, 1, \ldots, i, \ldots, m$. By specifying

$$g(Y = i; p_k) = \binom{m}{i} p_k^i (1 - p_k)^{(m-i)} \tag{16}$$

and inserting equation (16) in equation (3), the model of finite binomial mixtures is obtained. In addition to the latent class probabilities this model contains only one parameter per latent class, namely p_k, the success probability within the kth latent class.

To illustrate the usefulness of binomial mixture modeling in developmental psychology, consider the investigation of Tabor and Kendler (1981). Using only cross-sectional data, the authors tried to determine if the probabilities of solving Piagetian class inclusion problems and of choosing optional reversal shifts in Kendlerian concept learning tasks grow continuously or discontinuously in children. Discontinuous growth would be expected if Piaget's conceptions hold true; continuous growth would be expected if development of cognitive abilities is gradual.

Tabor and Kendler (1981) followed the following lines of reasoning: If development of the underlying cognitive capability is discontinuous, the probability of solving a problem should be discretely distributed with two realizations, one for "capable" and another for "incapable" subjects. If development is gradual and continuous, on the other hand, the same probability should be continuously distributed. The authors assumed that the former model generates bimodal, and the latter unimodal, histograms of solved problems when several equally difficult problems measuring the same capability are presented to subjects. Hence, the authors assumed that the developmental models can be distinguished by means of the number of modes in empirical histograms. This assumption is problematic (Murphy, 1964), as we have already discussed.

For a reanalysis of their data we need statistical models that connect the empirical distribution of solved problems with latent distributions of success probabilities. Provided that m equally difficult problems measuring the same cognitive capability were answered in a stochastically independent fashion, the distribution of solved problems must be binomial for fixed values of the success probability p. Since there are two such probabilities according to the stage conception, namely, p_1 for "capable" and p_2 for "incapable" subjects, the resulting distribution across all subjects should be a two-component binomial mixture.

However, what kind of mixture should we expect if p is continuously distributed? This distribution of p should (a) be defined on the real interval $[0, 1]$, the only reasonable interval for probabilities; (b) have only a few free parameters, so that the resulting mixture distribution has degrees of freedom comparable to the two-component binomial mixture; and (c) should be flexible so that almost any realistic distribution of success probabilities can be approximated by this function. The beta distribution is perhaps the best candidate. Fortunately, although beta mixtures of binomial distributions are not finite mixtures, the resulting mixture distribution, called the *beta binomial distribution,* is well known in the statistical literature, and the problem of parameter estimation has been solved. The interested reader is referred to Griffiths (1973) and Keats and Lord (1962).

TABLE 14.3. Reanalysis of Optional Reversal Shift Frequencies from the California Sample of Tabor and Kendler (1981)

(a) *Frequencies of optional reversal shifts*

Number of reversal shifts	Age group		
	Kindergarten	Grade 1	Grade 3
0	3	1	1
1	5	4	10
2	4	20	13
3	3	14	11
4	1	1	5
Sum	16	40	40

(b) *Results for the two-component binomial mixture*

Statistic	Age group		
	Kindergarten	Grade 1	Grade 3
$\hat{p}$(capable)	0.402	1.000	0.061
$\hat{p}_1$	0.629	0.563	0.955
$\hat{p}_2$	0.256	—	0.530
χ^2	0.023	6.307[a]	0.817

(c) *Results for the beta binomial model*

Statistic	Age group		
	Kindergarten	Grade 1	Grade 3
No. of modes	1	1	1
mean(p)	0.406	0.563	0.556
variance(p)	0.033	0.000	0.007
χ^2	0.069	6.309[a]	1.193

[a] Pearson chi-square statistics significant at the level alpha = 0.05.

A comparative analysis of Tabor and Kendler's data using the program BINOMIX (for ML analyses of finite binomial as well as of beta binomial mixtures) yields interesting results. Let us first look at the optional reversal shift data. One sample, comprised of three subsamples (kindergarten, first, and third grade children), was investigated with four structurally equivalent versions of the Kendlerian concept learning task (cf. Kendler and Kendler, 1970). For each subject, the number of reversal shifts in the four tasks was counted. The frequency distributions for the three subsamples, as estimated from Figure 4 in Tabor and Kendler (1981), can be read in Table 14.3(a). The Pearson chi-square statistic $\chi^2 = \sum_{j=0}^{4} (f_j - e_j)^2 / e_j$ was used to evaluate the models, where f_j and e_j denote the empirical and the expected (fitted) frequency of j reversal shifts, respectively. This statistic has some advantages compared to the LR-χ^2 statistic in the case of small sample sizes (cf. Milligan, 1980). Table 14.3(b) shows very good fit indices for two-component binomial mixture models in two of the three subsamples. However, the parameter estimates are not very reasonable. First, the estimated proportion of "capable" children [i.e., p(capable)] does not grow monotonically with age. Second, the success probabilities for "capable" and "incapable" children (i.e., p_1 and p_2, respectively) are not stable across age groups. These probabilities should be constant across age groups if the tasks measure the same capability in all subjects. Moreover, the first-grade children seem to produce no binomial mixture data at all. In this sample, the best fitting model is the one-component binomial mixture, or equivalently, the beta mixture with zero variance. However, even for this pathological solution the fit is bad.

All in all, the beta binomial model yields more reasonable results for this task, since the means of the latent beta distributions tend to grow monotonically with age. The fit indices, although for kindergarten and third-grade children still good, seem to be a little bit worse for the beta binomial model. Notice, however, that the two-component binomial mixture contains three free parameters, whereas the beta binomial models contains only two. Thus, the Pearson statistic has 1 df in the former and 2 df in the latter case.

Concerning continuous versus discontinuous development, the results clearly contradict the qualitative model and support the quantitative model. Hence, we may conclude that mediational learning, which is presumed to be the capability underlying optional reversal shift behavior (cf. Kendler and Kendler, 1970), does develop gradually during ontogenesis.

TABLE 14.4.   Reanalysis of Class Inclusion Responses from the California Sample of Tabor and Kendler (1981).

(a) *Frequencies of correct class inclusion responses*

Number of correct responses	Age group		
	Kindergarten	Grade 1	Grade 3
0	13	33	22
1	1	3	1
2	0	0	3
3	1	2	3
4	1	2	11
Sum	16	40	40

(b) *Results for the two-component binomial mixture*

Statistic	Age group		
	Kindergarten	Grade 1	Grade 3
$\hat{p}$(capable)	0.126	0.101	0.428
$\hat{p}_1$	0.870	0.870	0.863
$\hat{p}_2$	0.018	0.021	0.010
χ^2	0.341	0.726	3.409

(c) *Results for the beta binomial model*

Statistic	Age group		
	Kindergarten	Grade 1	Grade 3
No. of modes	2	2	2
Mean(p)	0.123	0.107	0.365
Variance(p)	0.075	0.062	0.184
χ^2	0.974	2.072	1.685

A completely different picture emerges if we look at class inclusion responses. Two samples of children, a California sample comprising three subsamples, and a North Dakota sample comprising four subsamples, were investigated with four structurally equivalent but semantically different class inclusion tasks. For each subject, the number of correct class inclusion responses in the four tasks was counted. The frequency distributions for both samples, as estimated from Figures 3 and 4 of Tabor and Kendler (1981), are reprinted in Tables 14.4(a) and 14.5(a),

TABLE 14.5.   Reanalysis of Class Inclusion Responses from the North Dakota Sample of Tabor and Kendler (1981).

(a) *Frequencies of correct class inclusion responses*

Number of correct responses	Age group			
	Kindergarten	Grade 2	Grade 4	Grade 6
0	31	22	16	8
1	2	0	1	0
2	0	0	3	3
3	0	3	4	6
4	3	7	16	15
Sum	36	32	40	32

(b) *Results for the two-component binomial mixture*

Statistic	Age group			
	Kindergarten	Grade 2	Grade 4	Grade 6
$\hat{p}$(capable)	0.083	0.313	0.577	0.750
$\hat{p}_1$	1.000	0.925	0.889	0.875
$\hat{p}_2$	0.015	0.000	0.015	0.000
χ^2	0.049	0.483	3.543	1.686

(c) *Results for the beta binomial model*

Statistic	Age group			
	Kindergarten	Grade 2	Grade 4	Grade 6
No. of modes	2	2	2	2
Mean(p)	0.107	0.269	0.506	0.633
Variance(p)	0.079	0.172	0.195	0.152
χ^2	2.824	6.038[a]	1.873	4.776

[a] Pearson chi-square statistics significant at the level alpha = 0.05.

respectively. The results of two-component binomial mixture analyses applied to these data clearly suggest good or excellent fits throughout (Tables 14.4(b) and 14.5(b)). Moreover, the proportion of "capable" children tends to grow monotonically with age, and the within-class success probabilities p_1 and p_2, as expected are fairly stable across subsamples.

Fit indices for the beta binomial model tend to be clearly worse than those of the finite mixture model. The fact that all estimated latent beta distributions have two modes (at $p = 0$ and $p = 1$) does not fit well with the idea of homogeneous subsamples. Hence the Piagetian notion of a *qualitative* transition from the preoperational to the operational stage of cognitive development seems to be in accordance with the results obtained for class inclusion problems, at least in this study.

5.5.   Mixtures of Products of Binomial Distributions

If multiple qualitative ordered variables, say, several rating scales, are obtained in one study, a multivariate generalization of finite binomial mixtures is possible. The assumption of stochastic independence of binomially distributed manifest variables within latent classes leads directly to the "latent class model for rating data" as described by Rost (1985). Of course, only for more than two categories per manifest variable is this model essentially new. For the case of dichotomous variables this model is equivalent to the latent class model discussed earlier.

5.6.   Other Kinds of Finite Mixture Distributions

Several other kinds of univariate and multivariate mixture models of potential importance for developmental psychologists may be mentioned: lognormal mixtures for positive, real-valued variables such as response times, exponential mixtures for waiting times (e.g., time until first success), Poisson mixtures for counts (e.g., number of solved problems per hour), or geometric mixtures for trials until first success (cf. Everitt and Hand, 1981; Titterington et al., 1985). *Hybrid mixture models* for data sets combining quantitative, qualitative nonordered, and qualitative ordered variables are also conceivable. Assume, for example, that the within-class local distribution is multivariate normal for one subset of manifest variables, a product of multinominal distributions for another subset, and a product of binomial distributions for a third. The local distribution for all variables combined could then be obtained by means of the product rule, for instance.

Hybrid finite mixture models have two advantages. First, they allow the analysis of data sets comprised of variables with heterogeneous scale properties within the framework of a common mixture model. Second, they allow the generalization of latent class analysis across several manifest groups (Clogg and Goodman, 1984, 1985) to other kinds of finite mixtures. The procedure will be essentially the same as described by Clogg and Goodman (1984, 1985): The manifest variable that indicates group membership must be defined as an exact indicator of one latent variable. This is done by restricting the category probabilities for this manifest variable to zero or one, respectively, given a certain latent class. Then parameter constraints across groups *for all types of variables* can be implemented.

6. Computer Programs for Finite Mixture Analyses

What kind of computer programs allow finite mixture analyses to be performed in a fast, flexible, and user-friendly way? Unfortunately, all available computer programs are limited to a fairly small range of finite mixture models. Hence, no generally applicable program can be recommended.

6.1. Programs for Univariate Mixtures

At present the program MIX 2.3 (Macdonald and Green, 1988)[1] seems to be the most comfortable for univariate mixture analyses. It interactively performs parameter estimation and evaluation for normal, lognormal, and gamma (including exponential) distribution mixtures. It is based on ML estimation for grouped (histogram) data, but does not use the EM algorithm. Instead, either quasi-Newton optimization or Nelder-Mead simplex optimization may be selected by the user. As mentioned earlier, this may result in problems in the event of "ugly" likelihood surfaces. Fortunately, the user's guide to MIX 2.3 (Macdonald and Green, 1988) gives much useful advice. One of the nice features of MIX 2.3 is the option of high-resolution graphics output. This option allows the application of informal graphical evaluation methods as well as the search for good starting values.

From the univariate mixture models discussed earlier, finite binomial mixtures and beta binomial distributions cannot be estimated and evaluated by means of MIX 2.3. The program BINOMIX, written and noncommercially distributed by this author, allows this. BINOMIX computes parameter estimates for finite binomial mixtures (using the EM

---

[1] The program MIX 2.3 is distributed on a commercial basis by Ichthus Data Systems, 59 Arkell Street, Hamilton, Ontario, Canada L8S 1N6.

algorithm) as well as for the beta binomial distribution (using a simple iterative scheme of ML estimation). Two different goodness-of-fit indices (i.e., LR-χ^2 and Pearson χ^2) are computed. Beyond that, semigraphic plots of the empirical histogram and the estimated beta density function are offered.

6.2. Programs for Multivariate Mixtures

While programs for multivariate normal mixtures may exist in several research groups, only a few are published. Wolfe's (1970, 1978) program exists in two versions: NORMAP and NORMIX, analyzing finite multivariate normal mixtures with and without the restriction of a common covariance matrix, respectively.[2] A formal LR-χ^2 goodness-of-fit test is offered by the program; however, see the remarks regarding this test in the section on problems of goodness-of-fit testing. Recently, McLachlan and Basford (1988) published another series of multivariate normal mixture programs based on the EM algorithm.

As far as LCA is concerned, the user has the choice between a variety of programs. If several LCA models differing in the number of latent classes must be analyzed for one data set, the program LCA from Rost and Sönnichsen (1982) has some advantages. However, a check of local identifiability of the model is missing in this program. Since no general answer can be given to the question of identifiability of LCA models, this option is quite important. The perhaps most famous LCA program, MLLSA, by Clogg (1977), provides this check. MLLSA can handle two kinds of parameter constraints, namely, equality constraints and fixation of parameter values. These features suffice to solve parameter estimation problems for many models of interest to developmental psychologists (Rindskopf, 1987). Among these are simultaneous latent class models for several groups (cf. Clogg and Goodman, 1984, 1985).

Formann's (1984) program LCALIN is based on a more general model that allows all varieties of linear constraints on logistically transformed parameters of traditional LCA. This model, called the *linear logistic latent class model*, has been investigated by Formann (1982, 1985). While LCA and MLLSA are based on the EM algorithm, the program LCALIN uses Newton-Raphson optimization or the scoring algorithm of ML estimation, with or without the gradient method applied first.

---

[2] According to Wolfe (1970, p. 344), NORMAP and NORMIX are distributed on a commercial basis by Focus Control Data Corporation, 3145 Porter Drive, Palo Alto, California 94304, USA.

Another program, Haberman's LAT (1979, pp. 586–596), can solve linear logistic LCA problems and more, but is not very user-friendly. A comparison of the different LCA programs and the underlying model variations is presented by Langeheine (1988). The source codes of LCALIN and LAT are reprinted in the appendixes of Formann (1984) and Haberman (1979), while the LCA and MLLSA source codes are distributed by the authors.

Researchers interested in applications of Rost's (1985) latent class model for rating data should use the Rost and Sönnichsen (1983) program, LCABIN, which performs goodness-of-fit tests as well as ML parameter estimation on the basis of the EM algorithm. The program is noncommercially distributed by the authors.

References

Aitkin, M., Anderson, D., and Hinde, J. (1981). "Statistical modeling of data on teaching styles," *Journal of the Royal Statistical Society, A.* **144,** 419–461.

Aitkin, M., and Wilson, G. T. (1980). "Mixture models, outliers and the EM algorithm," *Technometrics.* **22,** 325–331.

Andres, J., and Haessner, A. (1984). "Some problems concerning the chi-square goodness-of-fit test in LISREL," *Bonner Methodenberichte.* **1**(2) 1–27.

Bamber, B., and Van Santen, J. P. H. (1985). "How many parameters can a model have and still be testable?" *Journal of Mathematical Psychology.* **29,** 443–473.

Behboodian, J. (1970). "On the modes of a mixture of two normal distributions," *Technometrics.* **12,** 131–139.

Bergan, J. R. (1988). "Latent variable techniques for measuring development." In R. Langeheine and J. Rost (Eds.), *Latent Trait and Latent Class Models* (pp. 233–261). New York: Plenum.

Binder, D. A. (1978). "Comment on 'Estimating mixtures of normal distributions and switching regressions'," *Journal of the American Statistical Association.* **73,** 746–747.

Brandtstädter, J., and von Eye, A. (1982). "Aussagenlogische Analyse von Kontingenztafeln: I. Methodologische Vorüberlegungen," *Trierer Psychologische Berichte.* **9**(6) 1–18.

Campbell, D. T., and Fiske, D. W. (1959). "Convergent and discriminant validation by the multitrait-multimethod matrix," *Psychological Bulletin.* **56,** 81–105.

Carnap, R. (1936/1937). "Testability and meaning," *Philosophy of Science.* **3,** 418–471, **4,** 1–40.

Clogg, C. C. (1977). *Unrestricted and restricted maximum likelihood latent structure analysis: A manual for users.* Working Paper No. 1977-09. University Park, Pa.: Population Issues Research Office, Pennsylvania State University.

Clogg, C. C., and Goodman, L. A. (1984). "Latent structure analysis of a set of multidimensional contingency tables," *Journal of the American Statistical Association.* **79,** 762–771.

Clogg, C. C., and Goodman, L. A. (1985). "Simultaneous latent structure analysis in several groups." In N. W. Tuma (Ed.), *Sociological methodology.* San Francisco: Jossey-Bass.

Day, N. E. (1969). "Estimating the components of a mixture of normal distributions," *Biometrika.* **56,** 463–474.

Dempster, A. P., Laird, N. M., and Rubin, D. B. (1977). "Maximum likelihood estimation from incomplete data via the EM-algorithm," *Journal of the Royal Statistical Society, B,* **39,** 1–22.

Duda, R. O. and Hart, P. E. (1973). *Pattern classification and scene analysis.* New York: Wiley.

Erdfelder, E. (1988). "The empirical evaluation of deterministic developmental theories," *Berichte aus dem Psychologischen Institut der Universität Bonn.* **14**(3) 1–74.

Essler, W. K., and Trapp, R. (1978). "Some ways of operationally introducing dispositional predicates with regard to scientific and ordinary practice." In R. Tuomela (Ed.), *Dispositions* (pp. 109–134). Dordrecht: Reidel.

Everitt, B. S. (1981). "A Monte Carlo investigation of the likelihood ratio test for the number of components in a mixture of normal distributions," *Multivariate Behavioral Research.* **16,** 171–180.

Everitt, B. S., and Hand, D. J. (1981). *Finite mixture distributions.* London: Chapman and Hall.

Formann, A. K. (1982). "Linear logistic latent class analysis," *Biometrical Journal.* **24,** 171–190.

Formann, A. K. (1984). *Die Latent-Class-Analyse. Einführung in Theorie und Anwendung.* Weinheim W. Germany: Beltz.

Formann, A. K. (1985). "Constrained latent class models: Theory and applications," *British Journal of Mathematical and Statistical Psychology.* **38,** 87–111.

Gagné, R. (1965). *The conditions of learning.* New York: Holt, Rinehart, and Winston.

Griffiths, D. A. (1973). "Maximum likelihood estimation for the beta-binomial distribution and an application to the household distribution of the total number of cases of a disease," *Biometrics.* **29,** 637–648.

Guttman, L. (1981). "What is not what in theory construction." In I. Borg (Ed.), *Multidimensional data representations: when and why.* Ann Arbor, Mich.: Mathesis Press.

Haberman, S. J. (1979). *Analysis of qualitative data.* Vol. 2, *New developments.* New York: Academic Press.

Hasselblad, V. (1966). "Estimation of parameters for a mixture of normal distributions," *Technometrics*. **8**, 431–444.

Hasselblad, V. (1969). "Estimation of finite mixtures of distributions from the exponential family," *Journal of the American Statistical Association*. **64**, 1459–1471.

Hathaway, R. J. (1985). "A constrained formulation of maximum-likelihood estimation for normal mixture distribution," *Annals of Statistics*. **13**, 795–800.

Isaac, D. J., and O'Connor, B. M. (1969). "Experimental treatment of a discontinuity theory of psychological development," *Human Relations*. **22**, 427–455.

Johnson, N. L. and Kotz, S. (1972). *Distributions in statistics: multivariate distributions*. New York: Wiley.

Keats, J. A., and Lord, F. M. (1962). "A theoretical distribution for mental test scores," *Psychometrika*. **27**, 59–72.

Kendler, T., and Kendler, H. (1970). "An ontogeny of optimal shift behavior," *Child Development*. **41**, 1–27.

Kohlberg, L. (1976). "Moral stages and moralization: The cognitive-developmental approach." In T. Lickona (Ed.), *Moral development and behavior*. (pp. 31–53). New York: Holt, Rinehart, and Winston.

Langeheine, R. (1988). "New developments in latent class theory." In R. Langeheine and J. Rost (Eds.), *Latent Trait and Latent Class Models* (pp. 77–108). New York: Plenum.

Macdonald, P. D. M., and Green, P. E. J. (1988). *Users guide to program MIX: an interactive program for fitting mixtures of distributions*. Hamilton: Ichthus Data Systems.

McLachlan, G. J., and Basford, K. E. (1988). *Mixture models. Inference and applications to clustering*. New York: Dekker.

Milligan, G. W. (1980). "Factors that effect type I and type II error rates in the analysis of multidimensional contingency tables," *Psychological Bulletin*. **87**, 238–244.

Murphy, E. A. (1964). "One cause? Many causes? The argument from the bimodal distribution," *Journal of Chronic Disease*. **17**, 301–329.

Piaget, J. (1963). *The psychology of intelligence*. Paterson, N.J.: Littlefield.

Pinard, A., and Laurendeau, M. (1969). "'Stage' in Piaget's cognitive-developmental theory: exegesis of a concept." In D. Elkind and J. H. Flavell (Eds.), *Studies in cognitive development: essays in honor of Piaget* (pp. 121–163). New York: Oxford University Press.

Rindskopf, D. (1987). "Using latent class analysis to test developmental models," *Developmental Review*. **7**, 66–85.

Rost, J. (1985). "A latent class model for rating data," *Psychometrika*. **50**, 37–49.

Rost, J., and Sönnichsen, H. (1982). Die Analyse latenter Klassen: Eine Programmbeschreibung. *IPN Kurzberichte, 25.* Kiel, W. Germany: Institut für die Pädagogik der Naturwissenschaften.

Rost, J., and Sönnichsen, H. (1983). Probabilistische Testmodelle für Ratingdaten—Beschreibung der Computerprogramme RASBIN und LCABIN. *IPN Kurzberichte, 28.* Kiel, W. Germany: Institut für die Pädagogik der Naturwissenschaften.

Tabor, L. E., and Kendler, T. S. (1981). "Testing for developmental continuity or discontinuity: Class inclusion and reversal shifts," *Developmental Review.* **1,** 330–343.

Teicher, H. (1961). "Identifiability of mixtures," *Annals of Mathematical Statistics.* **32,** 242–248.

Teicher, H. (1963). "Identifiability of finite mixtures," *Annals of Mathematical Statistics.* **34,** 1260–1269.

Teicher, H. (1967). "Identifiability of mixtures of product measures," *Annals of Mathematical Statistics.* **38,** 1300–1302.

Thomas, H. (1987). "Modeling X-linked mediated development: development of sex differences in the service of a simple model." In J. Bisanz, C. J. Brainerd, and R. Kail (Eds.), *Formal methods in developmental psychology: progress in cognitive development research* (pp. 193–215). New York: Springer.

Titterington, D. M., Smith, A. F. M., and Makov, U. E. (1985). *Statistical analysis of finite mixture distributions.* Chichester, New York: Wiley.

von Eye, A., and Brandtstädter, J. (1988). "Application of prediction analysis to cross classifications of ordinal data," *Biometrical Journal.* **30,** 651–665.

Wilkening, F., Becker, J., and Trabasso, T. (1980). *Information integration by children.* Hillsdale, N.J.: Lawrence Erlbaum.

Wilks, S. S. (1938). "The large sample distribution of the likelihood ratio for testing composite hypotheses," *Annals of Mathematical Statistics.* **9,** 60–62.

Wolfe, J. H. (1970). "Pattern clustering by multivariate mixture analysis," *Multivariate Behavioral Research.* **5,** 329–350.

Wolfe, J. H. (1978). "Comparative cluster analysis of patterns of vocational interest," *Multivariate Behavioral Research.* **13,** 33–44.

Yakowitz, S. J., and Spragins, J. D. (1968). "On the identifiability of finite mixtures," *Annals of Mathematical Statistics.* **39,** 209–214.

Chapter 15   Prediction Analysis

KATHRYN A. SZABAT

Department of Management
LaSalle University
Philadelphia, Pennsylvania

Abstract

The developmental psychologist is often faced with investigations dealing with behavioral and/or educational development that concern variables measured in terms of discrete categories. The basic research design for such studies involves the cross-classification of each subject with respect to the relevant qualitative variables. Within this cross-classification framework, data analyses can be undertaken in terms of various measures and tests of association to address questions pertaining to the nature and extent of the relationship among variables. This chapter focuses on one data analysis tool, prediction analysis, as developed by Hildebrand, Laing, and Rosenthal (1977) (Material used by permission of John Wiley & Sons, Inc.).

Prediction analysis is a method for predicting empirical events based on specification of predicted relations among variables and includes techniques for both stating and evaluating those scientific predictions. This chapter presents the basic methodology employed in prediction analysis and highlights its features in comparison to the more conventional tests and measures used in the analysis of cross-classifications.

Prediction analysis is a general approach to the analysis of cross-classified data. It is a method for predicting empirical events based on specification of predicted relations among variables, and includes techniques for both stating and evaluating those scientific predictions.

Prediction analysis can be applied to theories predicting relations among quantitative variables (variables measured by ratio or interval scales), or to theories predicting relations among qualitative variables (variables measured by nominal or ordinal scales).

This chapter focuses on prediction analysis of cross-classifications of qualitative variables as developed by Hildebrand, Laing, and Rosenthal (1977). Specifically, it discusses prediction analysis as a data analysis tool for the developmental psychologist.

1. Prediction Analysis and Developmental Research

The developmental psychologist is often faced with longitudinal investigations dealing with behavioral and/or educational development that concern change for variables that are measured in terms of discrete categories (Nesselroade and Baltes, 1979). Such studies involve the classification of each subject with respect to several successive time points. In addition, cross-sectional studies often involve classifying subjects according to several categorical variables simultaneously at one time point. Thus, the basic research design for such studies involves the cross-classification of each subject with respect to relevant qualitative variables within the appropriate time frame.

Examples of categorical (qualitative) variables for these studies include measures of child competencies in task performance, verbalization patterns, program status, self-concept, health status, parent-child interaction, and developmental stages (Nesselroade and Baltes, 1979).

1.1. Hypotheses Involving Measures and Tests of Association

Within the cross-classification framework, data analyses can be undertaken in terms of various measures and tests of association to address questions pertaining to the nature and extent of the relationship among two or more categorical variables (also termed *attributes*).

Hypotheses can be directed at relations among attributes at a given time point and at the extent to which those relations change across over time. Hypotheses can be formulated about the extent to which individual subjects are classified into the same category for each attribute (such as developmental stages). Hypotheses can be developed to explore the joint agreement of several attributes as well as the pairwise agreement of two selected attributes. And, hypotheses concerning patterns of development such as synchronous convergent, convergent decalage, divergent decalage, and reciprocal interaction can be investigated (Nesselroade and Baltes, 1979).

1.2. Application of Prediction Analysis

Prediction analysis gives the developmental psychologist a research method for evaluating or selecting scientific predictions through analysis of cross-classified data. The scientific predictions are based on hypothesized relations among variables. Predictions (or propositions) can be stated a priori (before the data are analyzed) or selected ex post (after analysis of the data), and can be formulated as either degree-1 or degree-2 propositions. According to Hildebrand et al. (1977), "A degree-1 proposition makes a prediction for each observation taken one at a time" (p. 11). For example, predict level II on task B for subjects who have achieved level II on task A. "A degree-2 proposition makes predictions about the relative comparison of two observations" (p. 11) (Hildebrand et al. (1977) material used by permission of John Wiley & Sons, Inc.). For example, if the first subject in a pair of observations has achieved a higher level on task A than the other, predict that the first achieved a higher level on task B.

Prediction analysis addresses two questions:

1. What is the nature of the relation?
2. How strong is the relation?

That is, prediction analysis is concerned with predicting a type of relation and then measuring the success of that prediction. The method involves "prediction logic," expressing predictions that relate qualitative variables and the "del" measure used in analyzing data from the perspective of the prediction of interest.

1.3. Data Examples

To illustrate the use of prediction analysis, we use the following data examples.

Example 15.1. Table 15.1 reproduces a data set presented by Casey (1986) in her study of selective attention abilities. The data can be used to evaluate the prediction of agreement between the levels of the two

TABLE 15.1. Level of Performance on Two Tasks: Mirror-Image and Shape-Detail[a]

		Mirror-image level		
		1	2	
Shape-detail level	1	9	1	10
	2	3	7	10
		12	8	20

[a] Casey, M. B., Individual Differences in Selective Attention Among Prereaders. A Key to Mirror-Image confusions. Developmental Psychology, 1986, 22, 58–66. Copyright 1986 by American Psychological Association. Reprinted by Permission of the Author.

tasks: mirror-image and shape-detail. The levels of mirror-image and shape-detail are coded as follows: 1 = nonlearners (low), 2 = instructed learner (medium).

According to Kinsbourne's maturational theory, the shape-detail problem would be predicted to show equivalent-level effects with the mirror-image problem. Casey reports that, based on the Fisher test of exact probabilities, a significant relation was found when instructed

TABLE 15.2. Stages of Performance on Two Tasks: Seriation of Length and Inclusion/Length

		Seriation of length			
		I	II	III	
Inclusion length	I	14	15	19	48
	II	0	5	20	25
	III	0	2	26	28
		14	22	65	101

[a] Jamison, W., Developmental Inter-Relationships Among Concrete Operational Tasks: An Investigation of Piaget's Stage Concept. Journal of Experimental Child Psychology, 1977, 235–253. Copyright 1977 by Academic Press. Reprinted by Permission of Author.

TABLE 15.3.   Attribute Status at Time 1

			A_1 Attribute 1		
			1 Not present	2 Present	
A_2 Attribute 2	1	Not present	180	33	213
	2	Present	18	123	141
			198	156	354

learners (level 2) were compared with nonlearners (level 1) with respect to the two problems ($p = .02$). In addition, the kappa estimate for assessing the degree of subgroup agreement between levels 2 and 1 was .60 ($p = .008$).

Example 15.2.   Table 15.2 reproduces a data set presented by Jamison (1977) in his study of developmental interrelationships among concrete operational tasks. The data can be used to investigate the pattern of development known as *divergent decalage*, which states that development on one task (A) clearly takes precedence over development on the second task (B).

Jamison uses the chi-square goodness-of-fit procedure to evaluate the fit of the divergent decalage model to the cross-classification. The chi-square value was reported to be 4.78 ($p > .05$), indicating a good fit.

Example 15.3 and 15.4.   Tables 15.3 and 15.4 reproduce a data set presented by Landis and Koch (1979) in a discussion of categorical data analysis in longitudinal studies. The data can be used to compare the difference between the agreement of two attributes, A_1 and A_2, at two different time points, T_1 and T_2.

For each time period, Landis and Koch (1979) report log cross-product ratio, Yule's Q, and kappa (K) values, measuring the relation between the two attributes. At time T_1, $\Delta = 3.62$, $Q = .95$, and K = .70. At T_2, $\Delta = 2.97$, $Q = .90$, and K = .56. In addition, linearized minimum modified chi-square tests were used to test for differences between the two time periods with respect to each of the measures of association. The difference in kappa values was the only significant result.

TABLE 15.4.   Attribute Status at Time 2

			A_1 Attribute 1		
			1	2	
			Not present	Present	
A_2	1	Not present	57	39	96
Attribute 2	2	Present	18	240	258
			75	279	354

2.   Prediction Analysis for Bivariate Predictions: Degree-1

The initial focus of the discussion on prediction analysis is on measuring prediction success of degree-1, bivariate propositions in a population. The population approach is useful in explaining the methods of prediction analysis; prediction analysis of bivariate sample data is discussed in Section 5.4

2.1.   Prediction Logic

Let Y and X be two variables, independently measured, specified as the finite sets of states $(y_1, \ldots, y_R)$ and $(x_1, \ldots, x_C)$, where R and $C \geq 0$.

Let the domain of a prediction logic proposition include all events in the cross-classification $Y \times X$.

The predictions for observations having state X may be written in the form

$$P_{.j} : x_j \rightarrow S(x_j),$$

where the set of predicted ("success") states is a set of Y states. The proposition $P_{.j}$ is read, "if x_j, predict $S(x_j)$."

The event (y_i, x_j) belongs to the set of error events defining $P_{.j}$, that is,

$$\xi_{.j} = \{(y_i, x_j) \mid y_i \notin S(x_j)\}$$

if and only if that event would falsify the statement that X_j implies $S(X_j)$.

If we consider all states of X, the general prediction logic proposition for $Y \times X$ is stated as

$$P = \{P_{.j}\}$$

with error set

$$\xi_p = \bigcup_j \xi_{.j}$$

(Hildebrand et al., (1977) p. 31).

For Casey's study of selective attention abilities, the two variables, performance on mirror-image and performance on shape-detail, each have two states (level 1, level 2). Kinsbourne's maturational theory can be stated within the prediction logic framework as: "if the subject achieved level 1 on the mirror-image problem, predict he/she has achieved level 1 on the shape-detail problem; if the subject achieved level 2 on the mirror-image problem, predict he/she has achieved level 2 on the shape-detail problem." The diagonal cells of the cross-classification are the set of predicted or "success" states. The off-diagonal cells belong to the set of error events. That is, errors are committed when level 1 on shape-detail is predicted for subjects who have achieved level 1 on mirror-image but actually show level 2 on shape-detail, and when level 2 on shape-detail is predicted for subjects who have achieved level 2 on mirror-image but actually show level 1 on shape-detail.

The two variables in Jamison's study of developmental interrelationships among concrete operational tasks, stage on seriation of length and stage on inclusion/length, have three states each (stage I, stage II, and stage III). The divergent decalage model of development can be stated as: if the subject has reached stage I on the seriation of length task, predict he/she has reached stage I on the inclusion/length task; if the subject has reached stage II on the seriation of length task, predict he/she has reached stage I or II on the inclusion/length task; and if the subject has reached stage III on seriation of length, predict he/she has reached either stage I or stage II or stage III on inclusion/length. Only three of the cells in the 3×3 table (lower left triangle) are defined as error events. The rest are the "success" states.

Note, that in Table 15.1, only one "success" state is predicted for each given X state, but in Table 15.2 the prediction allows for more than one Y state to exist for a given X.

The proposition of attribute agreement in Table 15.3 can be phrased in the prediction logic framework in a similar way.

Once a theory is stated within the prediction logic framework, the prediction success of the proposition is measured.

2.2. Measure of Prediction Success

A proportionate reduction in error (PRE) model is used to define "del," the measure of prediction success. Define w_{ij} as the error cell indicator,

$$w_{ij} = \begin{cases} 1 & \text{if } (y_i, x_j) \in \xi_p \\ 0 & \text{if otherwise.} \end{cases}$$

Then the basic population measure of prediction success, ∇_p ("del"), for any admissible bivariate proposition P is defined as

$$\nabla_p = 1 - \frac{\sum_i \sum_j w_{ij} P_{ij}}{\sum_i \sum_j w_{ij} P_{i.} P_{.j}}, \tag{1}$$

where the cell probabilities of the $Y \times X$ cross-classification, P_{ij}, represent the probabilities that an observation belongs to states Y_i and X_j; and the marginal probabilities $P_{i.}$ and $P_{.j}$ indicate the probability that an observation lies in the ith row and jth column, respectively.

Comparison of equation (1) to the basic definition of a PRE measure shows that

$$\nabla_p = 1 - \frac{\text{Rule } K \text{ errors}}{\text{Rule } U \text{ errors}} = 1 - \frac{\text{Observed error rate}}{\text{Expected error rate}}.$$

The proposition P is applied to define both K and U errors. Rule K predicts a given set of states for all observations having a given X state. That is, it predicts the given set of Y states a given number of times (given number being equal to the number of observations having the given X state). Rule U makes the same prediction for the same number of observations without knowledge of the observations' actual X state. So, rule U predicts the same given set of Y states the same number of times, but now for observations selected at random. Therefore, ∇_p represents the PRE achieved in applying the prediction to observations with known X states over that which is expected when the prediction is applied to a random selection of observations with unknown X states.

Alternatively, "∇_p measures the PRE in the number of cases observed in the set of error cells for P relative to the number expected under statistical independence, given knowledge of the actual probability" (Hildebrand et al., (1977) p. 90. Material used by permission of John Wiley & Sons, Inc.).

The subscript p on ∇_p indicates that the ∇_p measures reflect a specific prediction. The subscript can be omitted when the prediction is evident. For convenience the subscript will be omitted from the ∇ notation for the rest of the chapter.

Recall Example 15.1. To measure the success of the prediction of agreement between the levels of the two tasks, mirror-image and shape-detail, specify the proposition P for the prediction of shape-detail levels: $1 \to 1$, $2 \to 2$. For the data of Table 15.1, with error events indicated as shaded cells, the measure of prediction success for the stated proposition P is calculated as follows:

$$\text{Rule } K \text{ error} = \left[\frac{1}{20} + \frac{3}{20}\right]$$

$$\text{Rule } U \text{ error} = \left[\left(\frac{10}{20}\right)\left(\frac{8}{20}\right) + \left(\frac{10}{20}\right)\left(\frac{12}{20}\right)\right] = .50,$$

$$\nabla = 1 - \frac{.20}{.50} = .60.$$

This value indicates that the number of errors made in applying the prediction P to randomly selected observations with unknown mirror-image levels is reduced by 60.0% when information about the actual level of performance on the mirror-image task is given.

The procedure, calculation, and interpretation is similar for Examples 15.2 and 15.3 and is summarized as follows:

Recall Example 15.2. To measure the success of the prediction based on the "divergent decalage" model, specify the proposition P for the prediction of levels of inclusion/length: I→I, II→I or II, III→I or II or III. For the data of Table 15.2, with error events indicated as shaded cells, the calculated measure of prediction success, ∇, is .851. A proportionate reduction in error of 85.1% is achieved in applying the divergent decalage proposition to subjects with known seriation of length performance stage over that which is expected when the prediction is applied to a random selection of subjects whose seriation of length performance stage is unknown.

Recall Example 15.3. To investigate the agreement between the two attributes A_1 and A_2 for the two time periods, T_1 and T_2, specify the proposition P for the prediction of the states of A_2 for each of the time periods as 1→1, 2→2. For the data of Table 15.3, with the error events indicated as shaded cells, $\nabla = .705$. For the data of Table 15.4, $\nabla = .563$. A greater reduction in prediction error is achieved at time 1.

2.3. Numerical Properties of ∇

The value of ∇ may vary in the range, $-\infty$ to 1: that is, $-\infty < \nabla \leq 1$. If no errors are committed under rule K (knowing the observations' X state), then $\nabla = 1$. If $0 \leq \nabla \leq 1$, then ∇ is the proportionate reduction in rule U errors obtained by applying the proposition under rule K. The ∇ value is negative if there are more observed errors than that which would be expected under statistical independence. This can be expected when a proposition is stated a priori simply because an a priori prediction is formulated according to a scientific (or research) theory and not according to any benchmark prediction. In ex post analysis, however, when predictions are formulated specifically to describe the data, ∇ should be positive.

If two variables are statistically independent of each other, it follows that $\nabla = 0$. The converse, however, does not hold. That is, $\nabla = 0$ does not imply that two variables are statistically independent. (An exception to the rule is the special cases of the 2×2 table.) The following example demonstrates the stated property. Suppose that for an $R \times C$ table, $P_{ij} \neq P_{i.} P_{.j}$ for all i, j. The variables thus are not statistically independent. Further, suppose that in the application of the ∇ measure, the total proportion of cases in the set of all error cells equals the total proportion expected in this set of cells under statistical independence. The rule K error rate is then equal to the rule U error rate. This leads to a ∇-value of 0. And so, the situation arises in which $\nabla = 0$, and yet the variables are statistically related.

2.4. Precision

When several competing theoretical predictions are compared to determine the dominance of a particular proposition, it is necessary to consider "precision" of a proposition in addition to the size of ∇.

For example, suppose for the data of Example 15.2, the developmental theory of reciprocal interaction is applied. For this theory, neither task takes precedence over the other. However, the degree of developmental divergence between tasks is limited. Subjects may reach level III simultaneously on the two tasks, but no subject will advance to level III on either task until he has attained level II on both (Jamison, 1977). The proposition for predicting levels of inclusion length is specified as $I \rightarrow I$ or II, $II \rightarrow I$ or II or III, $III \rightarrow II$ or III. The error events are the upper right and lower left corner cells of Table 15.2. The resulting value for prediction success is $\nabla = .454$. Comparison of this value to the prediction success value for the divergent decalage model ($\nabla = .851$) indicates that greater prediction success can be attributed to the latter model. Closer examination of the components of the ∇ measure for the divergent decalage model indicates that the prediction's empirical success, accuracy, has been achieved at the expense of prediction scope and precision.

A prediction's accuracy, which is defined as the extent to which prediction errors are minimized, is measured by the observed error (K) rate. For this model, the rule K error value is .0198, which is very small. A low error rate does not necessarily indicate that the prediction is successful. It may indicate that the prediction is close to "tautological," due to the limited scope of a prediction.

The *scope* of a proposition is defined as the proportion of the population for which nontautological predictions are made. The scope of a proposition increases with the proportion of nonvacuous predictions that are made. The scope of the component prediction, $P_{.j}$, is measured as

$$\text{Scope of } P_{.j} = \begin{cases} P_{.j} & \text{if } \xi_{.j} \neq \varnothing, \\ 0 & \text{otherwise.} \end{cases}$$

That is, each component prediction contributes to the effective scope of a proposition only if $P_{.j}$ identifies at least one error event. The overall scope of the proposition P is then calculated as the sum of the scope values for the component predictions. That is,

$$\text{Scope of } P = \sum_j \text{scope of } P_{.j}. \tag{2}$$

The scope of the proposition dictated by the divergent decalage model is .356. This indicates that the scope of the proposition extends to only 35.6% of the population. This, in turn, indicates that at most 35.6% of the observations can be predicted incorrectly.

The scope of the proposition generated by the reciprocal-interaction model, on the other hand, is .782, indicating a wider scope than that for the divergent decalage model.

Propositions with equal scope may differ with respect to the precision of the prediction. Propositions that are liberal in the specification of the error set tend to have more precision in prediction but, at the same time, have less accuracy.

Precision is determined, to a large extent, by the number of Y states that are predicted for given X states. Propositions that predict just one Y state for the given X states tend to be more precise than propositions that predict two or more states of Y. And this, of course, has an impact on accuracy.

The overall precision of a proposition is formally defined as

$$\sum_i \sum_j w_{ij} P_{i.} P_{.j}, \tag{3}$$

the rule U error rate. Larger values of U represent greater precision ($0 \le U \le 1$). Note that since the expected error rate is the precision of a proposition, ∇ can be interpreted as evaluating the observed error rate in relation to the prediction's precision.

Using equation (3) and the data of Table 15.2 (or by simply identifying the values of the rule U error rate in the calculation of ∇), we find the precision U for the divergent decalage and reciprocal-interaction models is .133 and .344, respectively. Both models have relatively poor precision. Even though the divergent decalage model has higher prediction success than the reciprocal-interaction model, the latter has higher precision.

The propositions stated in Example 15.1 and 15.3 have maximum scope value 1. The proposition of equivalent mirror-image/shape-detail performance levels in Example 15.1 has moderate precision, $U = .500$. The precision values of the proposition of equivalent attribute status in Example 15.3 are .288 and .368, at T_1 and T_2, respectively.

2.5. Dominance

One prediction dominates another when the prediction has a higher ∇ (prediction success) value and at least as great a U (prediction precision) value than the other, or a higher U-value and at least as great a ∇-value (Hildebrand et al., 1977, p. 94). Predictions that have a higher ∇-value will not always have at least as great a U-value, and a higher U will not always be associated with an at least as great a ∇-value. In either of these situations, the researcher must consider the trade-off, and he or she will be faced with the dilemma of deciding whether the data actually support the theoretical proposition.

2.6. Differential Weighting of Errors

Thus far, all error weights have been assumed to be 1, if the corresponding cell in the $R \times C$ table is a member of the set of error events, or 0, if a member of the set of "success" states. In some situations, however, some errors may be regarded as more important or more serious than others. For example, it is not unreasonable to postulate that observations further away from the main diagonal represent more severe errors than observations in the near-diagonal cells. In such cases, the error events can be differentially weighted. That is, each cell in the $R \times C$ table can be assigned some finite weight $w_{ij} \ge 0$.

2.7. Comparison with Measures and Tests of Association

The more conventional basic tools for analyzing relations among qualitative variables are tests and measures of association. Tests include the chi-square test of statistical independence and the chi-square goodness-of-fit test. Measures included the class of procedures related to the cross-product ratio (cpr), the measure of model prediction λ, the measure of proportional prediction τ, Cohen's kappa, and various measures of ordinal association: Goodman and Kruskal's γ, Somer's d_{yx}, Kim's d_{xy}, Kim's symmetric d and Kendall's τ. The various measures of ordinal association are all based on predictions about observation pairs and therefore will be discussed in the section on degree-2 analysis.

Each of the aforementioned measures was designed for a specific, somewhat limited, purpose and therefore does not meet all (although it may meet some) of the seven criteria for evaluating or designing a measure of prediction success, as proposed by Hildebrand et al. (1977b). According to Hildebrand et al., the model for measuring prediction success should

1. allow for custom tailoring to the specific a priori or ex post prediction under investigation;
2. produce similar results for logically equivalent propositions;
3. apply to predictions of any scope, precision, and differentiation, and note differences in these features along with accuracy in measuring prediction success;
4. allow a proportionate reduction in error interpretation;
5. provide for extension to multivariate analysis;
6. be relatively insensitive to minor changes in probabilities;
7. allow for the feasibility of accurate estimation with modest sample sizes.

In the discussion that follows, the conventional methods are related to these criteria.

The chi-square test of independence is used to determine whether there is any statistical relation between variables. Knowledge that two variables are not statistically unrelated (dependent) does not say anything about the form of the relation that exists. The chi-square statistic value is not a measure of the degree of the relation. Therefore, the basic problem with this chi-square test is that it focuses solely on statistical independence and is incapable of detecting the form of departure from independence.

Modifications of the chi-square statistic have been proposed as measures of association for determining the strength of association between two nominal variables (Capon, 1988). Pearson's phi (ϕ) and Cramer's V are nonproportionate reduction in error (non-PRE) measures. Because of this, interpretations are neither straightforward nor rigorous. Phi-square (ϕ^2) is based on a PRE measure. However, ϕ^2 gives no insight to the nature of a relation and, like chi-square, is fixed for a cross-tabulation regardless of what proposition is asserted (Hildebrand et al., 1977, p. 46).

The chi-square for goodness-of-fit is used to test the hypothesis that a specified set of probabilities P_{ij} determines the bivariate distribution. Thus, it allows for custom-designing. Despite this feature, the goodness-of-fit test shows the same deficiencies of the chi-square test of independence. In particular, it lacks a predictive nature. It cannot be given a PRE interpretation, and it focuses on predicting probabilities rather than events. Furthermore, goodness-of-fit tests are sensitive to small cell entries. Chi-square is undefined when observations occur for events that have a zero theoretical probability (Hildebrand et al., 1977, p. 51).

Several measures are based on the cross-product ratio for 2×2 tables, such as Yule's Q and Goodman's interaction analysis. Such measures have custom-designing features, but, like the goodness-of-fit approaches, they lack the desired predictive structure and are dictated by inferential rather than descriptive criteria. In addition, these measures are sensitive to small cell entries, a problem that also arises for the goodness-of-fit tests (Hildebrand et al., 1977, p. 51).

The goodness-of-fit and cpr approaches are not the only techniques that focus on predicting probabilities as opposed to events. Other techniques include a variety of log-linear methods. These models use a linear model approach to predicting probabilities. The emphasis is on modeling the nature and structure of a relation as opposed to simply testing for a dependence. Elliott (1988) called for a change in the modeling emphasis of log-linear analyses. In order to provide for more substantive interpretation of log-linear models, he proposed a means for interpreting higher-order interactions. Log-linear models, however, lack the necessary focus for measuring prediction success.

Guttman's λ and Goodman and Kruskal's τ are measures of association for nominal variables that come closest to satisfying Hildebrand et al.'s criteria. Both measures are concerned with event predictions and have PRE interpretations. The measure λ (using pure strategy prediction rules) represents the PRE committed in predicting each case lies in the modal Y-class when one moves from "not knowing the state of the independent variable" to "knowing the independent variable." The measure τ (using mixed strategy prediction rules) represents the PRE in selecting event predictions probabilistically in accordance with the observed distributions on the dependent variable when one moves from "knowing the marginals of the dependent variable" to "knowing each cases's location on the independent variable and the joint distribution." The major limitation of both λ and τ is that their prediction rules are totally inflexible and ex post. Thus, these measures lack the important feature of custom-designing.

Cohen originally proposed the measure kappa, K, as a measure of the degree of agreement between two nominal scales. To permit differential weighting of different types of disagreements between the two scales, Cohen later extended kappa to weighted kappa K_w. Hildebrand et al. noted the equivalence of K to ∇ for any square table, given the proposition P predicts the main-diagonal in the $R \times C$ cross-classification. Froman and Llabre (1985) demonstrated the formal equivalence of K_w to ∇, and proposed extensions to each statistic. But the demonstrated equivalences apply to a specific proposition stated a priori. Thus, as with λ and τ, K was not designed for custom-tailoring.

In summary, the more commonly used tests and measures of association fall short of satisfying all seven criteria for measuring prediction success, proposed by Hildebrand et al. On the other hand, ∇ was specifically developed with these criteria in mind.

3. Prediction Analysis for Bivariate Predictions: Degree-2

Degree-2 propositions, which make an event prediction for each pair of observations jointly, is likely relevant in research dealing with variables that involve a relative comparison of individuals, such as status. These interindividual comparisons are typically of the form, "one individual has greater, equal, or less status than another." Comparisons are not concerned with by how much they differ.

The prediction analysis methodology discussed previously can be employed for degree-2 propositions once a condensed form of the cross-classification of observation pairs is generated from the original $R \times C$ table of single observations.

3.1. The Condensed Ordinal Form

The cross-classification of the condensed form of two variables creates a 3×3 table with each variable condensed into three states: more, same,

TABLE 15.5.   Condensed Ordinal Form for Example 15.2

		Seriation of length			
		More	Same	Less	
	More	1562	1509	173	3244
Inclusion/ length	Same	913	1887	913	3713
	· Less	173	1509	1562	3244
		2648	4905	2648	10201

and less (see Table 15.5). In order to illustrate the conversion of an original $R \times C$ table of single observations to the condensed ordinal form necessary for degree-2 analysis, recall Example 15.2 and the data of Table 15.2. The following procedure was adapted from Hildebrand, Laing, and Rosenthal (1977a).

Step 1 involves the computation of the row and column totals. For seriation of length, there are 14 level I subjects who are tied on this scale. They can be paired in 14(14) = 196 different ways. Similarly, there will be 22(22) = 484 ties on level II and 65(65) = 4225 ties on level III. Across all seriation of length levels, there are 196 + 484 + 4225 = 4905 ties, and these are assigned to the category "same" of the condensed seriation of length variable. This number of ties can be used to compute the number of "more" and "less" pairs. Note, that these untied pairs cannot include self-pairs such as (S_1, S_1). If S_1 has a higher level of seriation of length than S_2, the pair (S_1, S_2) will be classified as "more," but the pair (S_2, S_1) will be classified as "less." Because of this symmetric counting of pairs, the number of pairs in the "more" and "less" categories must be equal. Now, the number of total pairs in the cross-classification is 101(101) = 10,201. The number of untied pairs, therefore, is 10,201 − 4905 = 5296. Because "more" and "less" have the same number of pairs, the number is calculated to be 5296/2 = 2648. This equals the column totals in the condensed ordinal form.

To get the inclusion/length (row) totals, repeat the previous procedure. There are 48(48) + 25(25) + 28(28) = 3713 tied pairs. By subtraction, there are 10,201 − 3713 = 6488 untied pairs. By symmetry, there are 6288/2 = 3244 pairs in the "more" and "less" categories.

The center entry in the condensed ordinal form, corresponding to a tie on both variables, is calculated next. There are 14 subjects that are level I on both seriation of length and inclusion/length. They are tied on both variables and can be paired in 14(14) ways. Squaring and summing over all cells of Table 15.2 in this manner, compute the total as $14(14) + 15(15) + 19(19) + 0(0) + 5(5) + 20(20) + 0(0) + 2(2) + 26(26) = 1887$.

Four more entries can be calculated from symmetry considerations. There are 4905 tied pairs on seriation of length and 1887 pairs tied on both variables. So there are $4905 - 1887 = 3018$ that are tied on seriation of length but not on inclusion/length. By symmetry, there are $3018/2 = 1509$ pairs that are "more" on inclusion/length but "same" on seriation of length, and 1509 that are "less" on inclusion/length but "same" on seriation of length.

Similarly, $3713 - 1887 = 1826$ pairs are tied on inclusion/length but untied on seriation of length. These are split equally, 913 being "more" on seriation of length and 913 being "less."

The four corner cells remain. The number of "more-more" pairs can be calculated first, and completion of the table will follow through simple calculation.

The 26 subjects showing level III on both seriation of length, and inclusion/length cannot be greater than any subjects in the last row and last column. But they do exhibit the "more-more" relation with all the other subjects in the table. To find these subjects, delete the third row and third column and add the rest. Schematically, the $(3, 3)$ cell constitutes $26(14 + 15 + 0 + 5) = 884$ pairs. There are 2 level III/level II subjects. These subjects exhibit the "more-more" relation with all subjects above and to the left in the table, constituting $2(0 + 14) = 28$ pairs. The first column can be ignored since there is no column to the left. Next, take the second row. The contributions are $20(14 + 15) = 580$ and $5(14) = 70$. The third row can be ignored since there is no row above it.

In total, therefore, there are $884 + 28 + 580 + 70 = 1562$ "more-more" pairs. If the pair (S_1, S_2) is "more-more," then (S_2, S_1) must be "less-less," so there are 1562 "less-less" pairs also. Therefore, $3244 - 1562 - 1509 = 173$ pairs "less" on seriation of length but "more" on inclusion/length, and 173 pairs are "more" on seriation of length but "less" on inclusion/length.

The condensed ordinal form is complete.

Hildebrand et al. (1977a) present degree-2 analysis in extensive form as well as in condensed form. Rules for computing the entries in the extensive form from the original $Y \times X$ cross-classification are given

(Hildebrand et al., 1977, p. 161). Once in extensive form, collapsing cells via summation then yields the condensed version.

3.2. The ∇ Measure for Degree-2 Propositions

Suppose that for Example 15.2, it is predicted that "if two subjects differ in attained level of seriation of length, the one who has attained a higher seriation level tends to have attained a higher level of inclusion/length." The degree-2 proposition in condensed ordinal form can be stated as: "more" → "more," "same" → "same," and "less" → "less."

For the data of Table 15.5, with error events indicated as shaded cells, and for the degree-1 calculation procedure, the measures of prediction success and precision are $\nabla = .229$ and $U = .659$, respectively. Thus, the propositon is somewhat precise but quite unsuccessful.

3.3. Comparison to Other Degree-2 Analyses

The more commonly known measures of association for degree-2 propositions are designed for ordinal variables: Goodman and Kruskal's γ, Somer's d_{yx}, Kim's d_{xy}, and Kim's symmetric d. Each of these can be given an a priori, PRE interpretation, and all are fixed prediction, typically of some form of monotonic relation between the variables. "If X increases, then Y tends to increase" is an example of such a prediction. These measures are thus limited in that they are unable to detect the presence of a nonmonotonic relation such as the S, J, or U shape (Hildebrand et al., 1977, p. 59).

The condensed ordinal form for applying these methods of ordinal analysis is typically displayed and represented as in Table 15.6. These $P(C)$ represents the probability of concordance, $P(D)$ the probability of discordance, $P(T_x)$ the probability of a tie on X, $P(T_{\bar{y}x})$ the probability of

TABLE 15.6.   The General Condensed Ordinal Form

	More	Same	Less	
More	$(1/2)P(C)$	$(1/2)P(T_{\bar{y}x})$	$(1/2)P(D)$	$(1/2)P(\bar{T}_y)$
Same	$(1/2)P(T_{y\bar{x}})$	$P(T_{yx})$	$(1/2)P(T_{y\bar{x}})$	$P(T_y)$
Less	$(1/2)P(D)$	$(1/2)P(T_{\bar{y}x})$	$(1/2)P(C)$	$(1/2)P(\bar{T}_y)$
	$(1/2)P(\bar{T}_x)$	$(1/2)P(T_x)$	$(1/2)P(\bar{T}_x)$	1.0

a tie on X but not on Y, $P(\bar{T}_x)$ the probability of not being tied on X, etc. The entries in the table can be calculated from the entries in an $R \times C$ table of probabilities via direct formulas.

Each of the conventional measures of ordinal association are functions of these probabilities. In each case, the numerator is the comparison of $P(C)$ to $P(D)$. The denominators, however, are different. In effect, the measures are identical except for how they handle ties.

Goodman and Kruskal's measure of ordinal association, λ, eliminates all ties before measuring association. This tends to exaggerate the degree of monotonicity in the original $Y \times X$ cross-classification.

Somer's and Kim's measure do not discard ties from the analysis. Somer's d_{yx} adjusts the difference between the probabilities of concordance and discordance on the basis of the independent variable. Kim's d_{xy} adjusts the difference based on the dependent variable. And Kim's symmetric d averages the two adjusting factors so that, as a result, d lies halfway between the two asymmetric measures d_{yx} and d_{xy}.

Using the notation in Table 15.6, Hildebrand et al. illustrate that the conventional measures are actually special cases of ∇. They identify the specific proposition P and associated error weights, such that ∇ is equivalent to the specific measure (see Hildebrand et al., p. 174).

In summary, even though these more conventional measures can be interpreted within the prediction logic framework, and despite their strengths, weaknesses are identified. Because the measures are a priori and are based on fixed predictions of some monotonic relation, they are incapable of addressing other possible scenarios that might be relevant in research. In contrast, ∇ provides a distinct measure for each distinct degree-2 proposition stated not only in condensed form but also in a more extended form. Furthermore, predictions can be stated a priori or ex post and as pure or mixed strategies. so the ∇ approach covers a wider area of research possibilities.

3.4. Degree-2 versus Degree-1 Analysis

Most prediction-related measures of association for nominal variables analyze the probability distribution in a cross-tabulation for single observations (degree-1). Most of the measures of association for ordinal variables are based on observation pairs (degree-2). Hildebrand et al. (1977a) argue that analysis of ordinal variables need not be restricted to pair predictions. In some research applications, a degree-1 prediction may be more appropriate than a degree-2 prediction. For example, the statement that one subject has a "higher degree of acceptance" than the other may have less predictive importance than the statement that the subject "strongly accepts." The first statement focuses on comparing observations to each other, while the second gives attention to the location of single observations in the ordered categories of the variable.

The nature of the prediction under investigation should dictate the degree analysis to apply. Now, ∇ was designed to permit custom-tailoring, whether the proposition is chosen for degree-1 or degree-2 analysis. The comparative nature of ordinal data is provided for, in the ∇ measure, by the comparison of conditional and unconditional probabilities, so pair predictions are not necessary (Hildebrand et al., 1977, p. 177).

4. Bivariate Statistical Inference

Until now, ∇ has been discussed as a population parameter. That is, a priori propositions have been evaluated with observations that represent an entire population. In reality, an investigator almost always works with a sample that is taken from a larger underlying population. As a result, the concern is one of inference; that is, using sample data to project something about the true value of ∇ in the population.

At this point, particular attention is given to the distinction between a priori and ex post propositions. The distinction becomes crucial in the statistical inference process because the statistical theory is not the same for both. In the discussion that follows, the prediction P is assumed to be specified a priori. Ex post analysis will be discussed in another section.

4.1. Estimates of ∇

Three sampling schemes cover the majority of research applications: (1) both sets of marginal proportions are known, and neither set of sample totals are fixed; (2) neither set of marginal proportions are known, and neither set of sample totals are fixed; and (3) only one set of marginal proportions is known, and the sample total for the corresponding variable is fixed. Given the specific sampling condition, the sample estimate $\hat{\nabla}$ of the population analogue is defined by substituting observed sample proportions for any unknown population probabilities. $\hat{\nabla}$ is the point estimate of the unknown population ∇ value, the single best guess. To evalute the probable precision of the estimator, it is necessary to establish its sampling distribution.

4.2. Sampling Distribution of ∇ Estimators

Given the sampling distribution of the ∇ estimators (any estimator, for that matter), it is possible to calculate the probabilities for each possible estimated value when sampling from a particular population and to calculate means, variances, and standard deviations. It is then possible to construct confidence intervals for the true value of ∇ and to calculate p-values for hypothesis testing.

The sampling distributions of ∇ estimators, given a priori prediction, are based on standard methods of asymptotic theories. The estimators are approximately normally distributed.

Under the second sampling scheme, which is probably the usual sampling situation, $\hat{\nabla}$ is asymptotically normal with mean ∇ and approximate variance,

$$\mathrm{Var}(\nabla) \cong (n-1)^{-1}\left[\sum_i \sum_j a_{ij}^2 P_{ij} - \left(\sum_i \sum_j a_{ij}P_{ij}\right)^2\right]$$

where

$$a_{ij} = U^{-1}[w_{ij} - B(\pi_{i.} + \pi_{.j})]$$

$$\pi_{i.} = \sum_j w_{ij}P_{.j}$$

$$\pi_{.j} = \sum_i w_{ij}P_{i.}$$

$$B = 1 - \nabla$$

(See Hildebrand et al., 1977, pp. 199–202, for other sampling scheme results.)

The variance value indicates the amount of probable deviation of $\hat{\nabla}$ from true ∇ in the population. Inspection of the formula indicates true population proportions to be known. However, sample proportions may be used in place of the population values without effecting the asymptotic distribution.

The following summarizes the calculation of the estimated variance for the data of Example 15.2. $\nabla = .851$, $U = .133$, $B = .149$, $\pi_{1.} = 0$, $\pi_{2.} = .139$, $\pi_{3.} = .357$, $\pi_{.1} = .525$, $\pi_{.2} = .277$, $\pi_{.3} = 0$, and $n = 101$, so the estimated variance equals .00974.

The variance calculations are rather tedious by hand, but the programming required to generate this value by computer is not difficult.

4.3. Confidence Intervals

Confidence intervals for the true value of ∇ in the population are based on a normal sampling distribution. The $100(1 - \alpha)\%$ confidence interval is

$$\hat{\nabla} \pm Z_{\alpha/2}[\text{est. var}(\hat{\nabla})]^{1/2}$$

where $Z_{\alpha/2}$ is the upper α percentage point from the standard normal table appropriate to the specified confidence level.

For Example 15.2, the 95% confidence interval is

$$.851 \pm 1.96(.0987)$$

$$.851 \pm .1934$$

That is, $.658 \le \nabla \le 1.00$, with 95% confidence.

4.4. Hypothesis Testing

A similar procedure can be applied in hypothesis testing. To test the hypothesis that the true value of ∇ is equal to some specified value ∇_0, as opposed to the one-sided alternative that $\nabla > \nabla_0$, use the test statistic

$$Z = \frac{\hat{\nabla} - \nabla_0}{[\text{est. var}(\hat{\nabla})]^{1/2}}$$

If Z_{α} is the critical one-tailed value for the specified significance level, then the alternative hypothesis is supported at that level if and only if $Z \ge Z_{\alpha}$.

Suppose for Example 15.2, the true value of ∇ is hypothesized to be greater than zero. The value of .851 for $\hat{\nabla}$ yields a test statistic of $Z = 8.62$, which is compared with the normal probability table value, 1.645. Therefore, conclude that the true value of ∇ is statistically greater than zero at the $\alpha = .05$ level.

4.5. Comparison of Two ∇-Values

The statistical inference methods just discussed can be extended to the problem of comparing two ∇-values. Methods are available for comparing two values that correspond to different predictions for the same data, or for two ∇-values that correspond to the same prediction for different data sets.

Hildebrand et al. (1977b) warn that these comparison methods may not produce insightful results. Because prediction analysis is many-faceted, sole comparison of two ∇-values may not yield appropriate conclusions. Prediction analysis evaluates not only the degree of error reduction (∇) but also the degree of precision (U) and the scientific plausibility of the prediction. A prediction that yields a large ∇-value may not be more important or valuable than a second prediction that yields a smaller value, because the latter may have much greater scope and precision. In the case where the same prediction rule is applied to two different data sets, the problem reduces more completely to a sole comparison of ∇-values. The scientific plausibility is most likely the same in both instances, and the precision measures should be the same, unless the marginals vary drastically.

The testing and estimation procedures for comparing two ∇-values is a direct extension to the previous methods.

4.5.1. Hypothesis Testing.

To compare estimated ∇-values for two different, independently gathered data sets, test the null hypothesis of equal ∇-values via the test statistic

$$Z = \frac{\hat{\nabla}_1 - \hat{\nabla}_2}{[\text{est. var}(\hat{\nabla}_1) + \text{est. var}(\hat{\nabla}_2)]^{1/2}},$$

where the individual variances are computed separately, using the variance formulas appropriate to the sample scheme (such as the formula for the second sampling scheme given previously).

The test statistic value is then compared, as before, with a normal table value.

To compare ∇-values for two predictions on the same data, test the null hypothesis of equal ∇-values via the test statistic

$$Z = \frac{\hat{\nabla}_1 - \hat{\nabla}_2}{[\text{est. var}(\hat{\nabla}_1 - \hat{\nabla}_2)]^{1/2}},$$

where

$$\text{var}(\hat{\nabla}_1 - \hat{\nabla}_2) = (n-1)^{-1} \left[\sum_i \sum_j (a_{ij}^{(1)} - a_{ij}^{(2)})^2 P_{ij} - \left(\sum_i \sum_j (a_{ij}^{(1)} - a_{ij}^{(2)}) P_{ij} \right)^2 \right].$$

The calculation for the estimate of var ($\hat{\nabla}_1 - \hat{\nabla}_2$) is similar (although a bit more complicated) to the typical variance calculation. The difference is that a_{ij} is replaced by $a_{ij}^{(1)} - a_{ij}^{(2)}$. As before, the resulting test statistic value is compared with a value taken from the normal distribution table.

4.5.2.   Confidence Intervals.   The $100(1 - \alpha)\%$ confidence interval for the difference in two ∇-values is

$$(\hat{\nabla}_1 - \hat{\nabla}_2) \pm Z_{\alpha/2}[\text{est. var}(\hat{\nabla}_1) + \text{est. var}(\hat{\nabla}_2)]^{1/2}$$

for the case corresponding to the same prediction for two different, independently gathered, data sets.

For the case corresponding to two different predictions on the same data, the $100(1 - \alpha)\%$ confidence interval is

$$(\hat{\nabla}_1 - \hat{\nabla}_2) \pm Z_{\alpha/2}[\text{est. var}(\hat{\nabla}_1 - \hat{\nabla}_2)]^{1/2}$$

4.6.   Adequacy of the Normal Approximation

Hildebrand et al. (1977b) report that the essential condition for the applicability of the normal approximation under sampling schemes 1 and 2 is that $5 \leq nU(1 - \nabla) \leq n - 5$, the standard condition for applying the normal distribution to the binomial. If this condition is met, a researcher may use the approximation, except when marginals are extremely skewed or when true values of ∇ are usually large.

In particular, for confidence intervals, a large U and modest ∇ indicate a good approximation. For testing, under the null hypothesis that true $\nabla = 0$, the rule of thumb becomes $nU \geq 5$. Note that this rule refers to the overall expected error rate in the set of all error cells as a whole, not to each error cell individually. As a result, there is no gain in collapsing categories of large $R \times C$ cross-classifications.

The same principles most likely apply under sampling scheme 3, but there is no direct proof to support the claim. A more detailed discussion of the adequacy of the normal approximation is given in Hildebrand et al. (1977b).

4.7.   Continuity Correction

The normal approximation can be improved by a continuity correction. For estimation, the corrected $100(1 - \alpha)\%$ confidence interval is

$$\hat{\nabla}^+ - Z_{\alpha/2}[\text{est. var}(\hat{\nabla})]^{1/2} \leq \nabla \leq \hat{\nabla}^- + Z_{\alpha/2}[\text{est. var}(\hat{\nabla})]^{1/2},$$

where

$$\hat{\nabla}^+ = 1 - \frac{\hat{K} + D/n}{\hat{U}},$$

$$\hat{\nabla}^- = 1 - \frac{\hat{K} - D/n}{\hat{U}},$$

$$D = \begin{cases} 1.00 & \text{for a } 2 \times 2 \text{ table with diagonally opposed} \\ & \text{error cells or any logically} \\ & \text{equivalent table,} \\ .50. & \text{otherwise.} \end{cases}$$

The corrected expression for hypothesis testing is

$$Z = \frac{\hat{\nabla}^+ - \nabla_0}{[\text{est. var}(\hat{\nabla})]^{1/2}}.$$

Note, no corrections for continuity are necessary for calculating a variance of $\hat{\nabla}$ or in computing the point estimate $\hat{\nabla}$.

Recall Example 15.2. Application of the rules of thumb for determining the adequacy of the normal approximation indicate the quality of the normal approximation in the estimation technique could be improved with a continuity correction.

4.8. Ex Post Analysis

The importance of distinguishing between a priori and ex post proposition was previously noted. The statistical theory presented thus far applies solely to predictions stated before the data are analyzed. Therefore, the a priori statistical procedures developed do not apply to predictions selected after the analysis the data. Use of standard a priori methods in an ex post problem leads to erroneous probability statements and the possibility of claiming results not due to chance when they are indeed due purely to chance.

In addition to the lack of standard methods of statistical theory for ex post predictions, there is a serious methodological problem associated with ex post analysis. The problem is that of bias. The estimate of true ∇ is automatically biased upward; this follows because the proposition is, almost by definition, selected from the data to yield a large $\hat{\nabla}$-value.

Because a researcher may not always be in the position to apply a priori theory, some discussion should be directed to the concerns (bias, in particular) faced by the researcher in an ex post analysis.

4.8.1. Ex Post Selection Rules. Before discussing bias, attention is given to rules that dictate ex post selection of an "optimal" proposition.

Hildebrand et al. (1977b) consider a U-optimal criterion. In an ex post search for an optimal proposition, they propose to maximize ∇ subject to the attainment of a prespecified level of precision U. Recall, that U is the denominator of the ∇-measure that represents the boldness of the prediction made by a proposition in the population. In their selection procedure, error cells are chosen from the $R \times C$ table until the specified level of precision is met.

Bartlett's procedure (1974) prespecifies l, the minimum value a cell prediction success measure (∇_{ij}) can be to be included as an error cell. Then all cells with $\nabla_{ij} > l$ are considered error events. The approach is useful when there are a large number of X states, or when there is no or little a priori theory for the bivariate relation, or when a parsimonious ex post description is desired.

This author used a slightly different criterion. Instead of determining the number of errors via the U-optimal rule or via the Bartlett specification, the number of error cells to be included is specified at the onset (Szabat, 1982). For example, the researcher might specify the number of errors cells to be equal to one half the total number of cells in the $R \times C$ classification.

This selection rule allowed for an investigation of the properties of bias.

4.8.2. Bias. The author examined the question of how much bias is present in ex post selection of propositions via the sampling distribution of estimators that were defined by the aforementioned ex post selection rule. Theoretically derived expressions show that bias is a function of the expected value of asymptotically normal order statistics with parameters n, RC, T, and ∇; n is the sample size, RC the table size, T the specified number of errors cells used in defining an ex post proposition, and ∇ the true value of ∇ for the population. Established properties of bias are:

(a) For a given table size RC and specified number of error cells T, the bias of $\hat{\nabla}$ is a decreasing function of the sample size n.

(b) For a given sample size n and table size RC, the bias of $\hat{\nabla}$ decreases as the number of error cells T increases. Furthermore, bias decreases at a decreasing rate for $T \leq RC/2$.

(c) For a given number of error cells T and sample size n, the bias of $\hat{\nabla}$ increases as the table size RC increases for $T \leq RC/2$.

(d) For a given sample size n and number of error cells T as a fraction of RC, the bias increases as the table size RC increases.

(e) Bias decreases as true population ∇ increases.

(f) The sharper the distinction between cell del values, the faster the bias goes to zero.

These properties hold for all sampling cases. In addition, it is noted that bias decreases as the shift from known population marginals to unknown population marginals is made.

Table 15.7 reports theoretical bias values for three specified population dels under sampling scheme 1. Note the high bias value (.3917) when true del is 0 in the population.

In summary, both theoretical and simulated results indicate that selecting a proposition ex post can yield spuriously large values when true del is zero, particularly for small sample sizes, large tables, and few error cells. When true ∇ is positive, the problem is not as serious, especially if true ∇ is large. If a strong relation exists, a researcher can be reasonably confident that it will be assessed accurately.

The results clearly identify the dilemma faced by a researcher who

TABLE 15.7. Comparison of Theoretical Bias Values for Three Specifications of True Population Del ($S1$, $T = (1/2)RC$ or $(1/2)RC + 1$) (Reprinted with permission from John Wiley & Sons, Inc., Hildebrand, D., Laing, M., and Rosenthal, A. *Prediction analysis of cross classifications* (1977)).

Table size	Marginals[a]	Sample size	Bias		
			$\nabla = 0$	$\nabla = .20$	$\nabla = .70$
2×2	E	25	.2656	.1089	.0060
2×2	E	100	.1326	.0159	.0000
2×2	U	25	.2652	.1063	$-.0311$
2×2	U	100	.1326	.0136	$-.0380$
2×3	E	25	.3446	.1859	.0259
2×3	E	100	.1722	.0478	$-.0020$
2×3	U	25	.3416	.1861	$-.0080$
2×3	U	100	.1722	.0529	$-.0523$
3×3	E	25	.3917	.1866	.0781
3×3	E	100	.1958	.0571	.0180
3×3	U	25	.3917	.2386	.0940
3×3	U	100	.1958	.0785	.0194

[a] Marginal distribution specification: E = equal (uniformly distributed); U = unequal (evenly spaced across rows and across columns, therefore, mildly skewed)

obtains a moderate to large value for $\hat{\nabla}$. Is the large value a result of the fact that there is a strong relation in the population? Or is it actually a result of the selection bias induced by ex post methods?

4.8.3. Ex Post Statistical Inference. The aforementioned problems hinder the search for a satisfactory procedure for assessing an estimated ∇-value. Hildebrand et al. present an approach that could be used to test whether the positive $\hat{\nabla}$-value for an ex post proposition is attributable to chance. The statistical theory is closely related to the chi-square test. The test statistic is simply the square of the estimated ∇ divided by its estimated variance, and is compared to a chi-square table with $(R - 1) \times (C - 1)$ degrees of freedom. It is a two-tailed test that is unsatisfactory, since prediction analysis aims to specify propositions with positive values (Hildebrand et al., 1977, p. 222). Wasserman (1973) applied this chi-square test to the U^*-optimal selection rule and found the test to be very conservative for 3×5 tables and extremely so for 5×7 tables. Results, under the hypothesis of independence, showed only a .1% true rejection rate at a nominal 5% rate. Wasserman found that if a $\hat{\nabla}$-value is declared "ex post significant" by the chi-square test, it is very likely to be nonzero; yet a value falling short of significance by the chi-square test might be significant under a more exact, less conservative test. The chi-square test is not a practically effective test.

As an alternative, this author proposes a pseudo-Z-test (Szabat, p. 355). That is, use the normal curve to conduct tests of hypotheses. Normal curve methods require only the specification of the mean and variance that apply to the particular sampling distribution. The means and variances of the various sampling estimates, under the "given number of error cells" selection rule, are computable. The sampling distribution of the estimators is not claimed to be normal. The author simply proposes a more satisfactory method by which one can test a positive $\hat{\nabla}$-value for an ex post proposition. Simulation experiments compared actual rejection rates under the true hypotheses of independence, as against nominal 5% rate. The results are satisfactory (within simulation error) and indicate that the Z-test may be a more preferred test over the chi-square test proposed by Hildebrand et al.

In summary, procedures for ex post assessment of propositons are not entirely satisfactory. Uncertainties in an ex post approach and possible biases in selection of variables and in selection rules indicate that perhaps the only way to establish an ex post analysis is by confirmation in a later study. This is possible within the prediction analysis framework. A researcher can develop a prediction rule ex post in one analysis and then apply that prediction a priori in a replication.

5. Multivariate Analysis

This section discusses the extension of bivariate analysis to multivariate analysis. More specifically, it addresses the prediction of a single dependent variable when two predictor variables are used (trivariate case).

The intent is to present the basic framework for conducting multivariate analysis but not to cover all considerations of the analysis. For a more extensive discussion, see Hildebrand et al.

The focus here is on (1) how to state predictions based on two predictor variables and (2) how to measure the overall success of such predictions.

5.1.   Trivariate Data Example

Recall Jamison's study of developmental interrelationships among concrete operational tasks. Suppose that a third variable, seriation of area, is considered as a second predictor variable for predicting the dependent variable, inclusion/length. That is, now both stage of performance on seriation of length and stage of performance on seriation of area are used to predict a subject's stage of performance on inclusion/length. The data for this trivariate example is given in Table 15.8.

5.2.   Multivariate Prediction Logic

Denote the dependent variable as Y and the two predictor variables as X and W, with states $(y_1, \ldots, y_R)$, $(x_1, \ldots, x_c)$, and $(w_1, \ldots, w_s)$, respectively, where s is the number of strata in the three-way cross-classification. Create the Cartesian product of the two predictor variables. The Cartesian product represents all possible pairs of X and W states, and in effect creates a new variable V with $v_1 = (x_1 w_1), \ldots, v_{cs} = (x_c w_s)$. The original three-way table, Table 15.8, is then condensed into a two-way table, with columns representing the possible pairs of X and W states and rows representing the Y states. The condensed trivariate summarization for the data example is given in Table 15.9.

TABLE 15.8. Trivariate Cross-Classification of Seriation of Area, Seriation of Length, and Inclusion/Length

Seriation of area = I

Seriation of length

		I	II	III	
	I	13	3	0	16
Inclusion/ length	II	0	0	0	0
	III	0	1	1	1
		13	3	1	17

Seriation of area = II

Seriation of length

		I	II	III	
	I	1	12	13	26
Inclusion/ length	II	0	5	8	13
	III	0	1	15	16
		1	18	36	55

Seriation of area = III

Seriation of length

		I	II	III	
	I	0	0	6	6
Inclusion/ length	II	0	0	12	12
	III	0	1	10	11
		0	1	28	29

For multivariate prediction, predict some set of Y states for every possible pair of X and W states. Similar to bivariate analysis, the set of predicted Y states for the given $(X \& W)$ pair constitutes the set of "success" states, and corresponding cells are assigned weights of 0; the unpredicted Y states are members of the error set and are assigned weights of 1.

In effect, the multivariate problem has been transformed into an equivalent bivariate problem.

TABLE 15.9. Condensed Trivariate Form of Seriation of Area, Seriation of Length, and Inclusion/Length

		Seriation of length & seriation of area									
		I & I	I & II	I & III	II & I	II & II	II & III	III & I	III & II	III & III	
Inclusion/length	I	13	1	0	3	12	0	0	13	6	48
	II	0	0	0	0	5	0	0	8	12	25
	III	0	0	0	0	1	1	1	15	10	28
		13	1	0	3	18	1	1	36	28	101

To illustrate a trivariate proposition, return to the trivariate data example described earlier. Let the levels of seriation of length, seriation of area, and inclusion/length be denoted as $(X\text{I}, X\text{II}, X\text{III})$, $(W\text{I}, W\text{II}, W\text{III})$, and $(Y\text{I}, Y\text{II}, Y\text{III})$, respectively. A hypothetical proposition might be stated as follows:

$$X\text{I} \ \& \ W\text{I} \rightarrow Y\text{I}$$

$$X\text{I} \ \& \ W\text{II} \rightarrow Y\text{I}$$

$$X\text{I} \ \& \ W\text{III} \rightarrow Y\text{I}$$

$$X\text{II} \ \& \ W\text{I} \rightarrow Y\text{II}$$

$$X\text{II} \ \& \ W\text{II} \rightarrow Y\text{II}$$

$$X\text{II} \ \& \ W\text{III} \rightarrow Y\text{II}$$

$$X\text{III} \ \& \ W\text{I} \rightarrow Y\text{III}$$

$$X\text{III} \ \& \ W\text{II} \rightarrow Y\text{III}$$

$$X\text{III} \ \& \ W\text{III} \rightarrow Y\text{III}$$

The first prediction statement can be translated as follows: "if the subject performs at the stage I level for both seriation of length and seriation of area, predict that he or she performs at the stage I level for inclusion/length." Similar translations can be phrased for the remaining prediction statements.

5.3.   The Measure of Multivariate Prediction Success

The multivariate V measure for the trivariate proposition is defined as

$$\nabla_{P_{yxw}} = 1 - \frac{P_{yxw} \text{ error, given } (X \& W)}{P_{yxw} \text{ error, given random } (X \& W)}$$

$$= 1 - \frac{\sum_i \sum_j \sum_k w_{ijk} P_{ijk}}{\sum_i \sum_j \sum_k w_{ijk} P_{i..} P_{.jk}},$$

where

$P_{ijk} = $ the three-way joint probabilities of $Y \times X \times W$

$P_{i..} = \sum_j \sum_k P_{ijk} = $ row probabilities

$P_{.jk} = \sum_i P_{ijk} = $ column probabilities

taken from the condensed trivariate table.

For the trivariate proposition, and using the data in Table 15.9, with error events indicated as shaded cells, and the computational procedure as in bivariate analysis, the measure of prediction success is .216. A proportionate reduction in predictive error of 21.6% is achieved when the trivariate proposition is applied to observations whose joint seriation of length and seriation of area performance stages are known over that which is expected if the prediction is applied to a random selection of observations for whom joint performance stages are unknown.

In summary, the multivariate ∇ measure is simply the bivariate ∇ applied with a composite independent variable. Once the composite variable is defined, the techniques are the same as for a bivariate analysis.

The analysis need not end here, however.

Hildebrand et al. (1977b) extend the analysis to include discussion of the following topics: (1) measurement of partial success for analyzing the separate effects of the various predictor variables; (2) error accounting for exploring relations among the three ∇ measures: bivariate ∇, multivariate ∇, and average partial ∇; (3) statisticald inference; and (4) ex post analysis. See Hildebrand et al. (1977, p. 263).

6.   Summary

The developmental researcher is frequently faced with the task of evaluating established research theories about relationships among qualitative variables, such as Jamison in his study of developmental inter-relationships among concrete operational tasks. In addition, the situation could arise in which no a priori theory can be formulated, so the researcher is confronted with the problem of selecting a plausible theory about a relationship based on an analysis of collected data.

Prediction analysis is a viable, valuable data analysis tool for the developmental researcher in these researcher settings. It is based on the analysis of cross-classifications of qualitative data and is set within a prediction logic framework. Prediction analysis employs a prediction logic language that permits a precise statement of predictions and a statistical measure that measures the prediction success for each distinct prediction in the statement. The analysis can be applied to one-to-one and one-to-many predictions about variables, to bivariate and multivariate settings, to predictions stated a priori and selected ex post, to single observations and observation pairs, and to pure and mixed strategies. Thus, prediction analysis should cover a wide range of research possibilities.

This chapter has presented the basic methodology employed in prediction analysis and has highlighted its features in comparison to the more conventional tests and measures of association used in the analysis of cross-classifications. Interested readers are referred to Hildebrand, Laing, and Rosenthal's book *Prediction analysis of cross classifications* for complete, detailed coverage of the prediction analysis approach.

References

Bartlett, R. A. (1974). "Partition analysis of categorical data." Unpublished Doctoral Dissertation. Philadelphia: University of Pennsylvania.

Bishop, Y. M. M., Feinberg, S. E., and Holland, P. W. (1975). *Discrete multivariate analysis: theory and practice.* Cambridge, Mass.: MIT Press.

Capon, J. A. (1988). *Elementary statistics for the social sciences.* Belmont, Calif.: Wadsworth, 1988.

Casey, M. B. (1986). "Individual differences in selective attention among prereaders. A key to mirror-image confusions," *Developmental Psychology.* **22,** 58–66.

Elliott, G. C. (1988). "Interpreting higher order interactions in log-linear analysis," *Psychological Bulletin.* **103,** 121–130.

Froman, T., and Lawrence, J. H. (1980). "Application of prediction analysis to developmental priority," *Psychological Bulletin.* **87,** 136–146.

Froman, T., and Llabre, M. (1985). The equivalence of kappa and del," *Perceptual and Motor Skills.* **60,** 3–9.

Hildebrand, D., Laing, M., and Rosenthal, A. (1977a). "Analysis of ordinal data," Sage University Paper Series on Quantitative Application in the Social Sciences, 07–008. Beverly Hills and London: Sage.

Hildebrand, D. Laing, M., and Rosenthal, A. (1977b). *Prediction analysis of cross classifications.* New York: Wiley.

Jamison, W. (1977). "Developmental inter-relationships among concrete operational tasks: an investigation of Paiget's stage conception," *Journal of Experimental Child Psychology.* **24,** 235–253.

Landis, J. R., and Koch, G. G. (1979). "The analysis of categorical data in longitudinal studies of behavioral development." In J. Nesselroade and P. B. Bates (Eds.), *Longitudinal research in the study of behavior and development* (pp. 233–261). New York: Academic Press.

Nesselroade, J. R., and Baltes, P. B. (1979). *Longitudinal research in the study of behavior and development.* New York: Academic Press.

Szabat, K. (1982). "Ex post del." Unpublished Doctoral Dissertation, Philadelphia: University of Pennsylvania.

Wasserman, S. (1973). "An expost significance test for a pattern of association based on predictive logic." Unpublished M.S. Thesis. Philadelphia: University of Pennsylvania.

Configural Frequency Analysis of Longitudinal Multivariate Responses

Chapter 16

ALEXANDER VON EYE*

Department of Human Development and Family Studies
College of Health and Human Development
The Pennsylvania State Univeristy
University Park, Pennsylvania

Abstract

This chapter discusses methods for analysis of longitudinal multivariate responses with configural frequency analysis (CFA). The typical approach to analysis of longitudinal categorical data with log-linear models of CFA crosses all variables across all occasions. This often results in an unmanageable number of contingency table cells. Three methods of reducing the number of cells are discussed. The first approach is to analyze differences rather than untransformed variables. This leads to a reduction of a number of cells for short time series only. The second approach, more efficient for longer time series, is to approximate time-series data with orthogonal polynomials. The polynomial coefficients can then be categorized and analyzed with CFA. The third method relates adjacent vectors of observations to each other and analyzes the resulting correlation or distance coefficients. This method is most efficient in multivariate designs. The chapter gives data examples for each of the approaches.

* The author is indebted to Phil Wood, Connie James, Paul Games, and Chip Scialfa for helpful comments on earlier versions of this chapter.

1. Configural Frequency Analysis

Researchers may apply log-linear models to analyze the dependence structure of variables in a contingency table. The models estimate the logarithms of expected frequencies as linear functions of parameters that depict the variables' associations. Log-linear models typically consider associations among more than two variables. Log-linear modeling proceeds in two steps (see Clogg et al., this volume). First, based on the researcher's hypothesis regarding the association structure of variables, expected frequencies are estimated. Second, likelihood ratio or Pearson chi-square tests are applied to determine whether the deviations of expected frequencies based on the model from the observed frequencies can be considered statistically random.

Even if these deviations, or residuals, are not statistically significant, they are rarely zero. The analysis of these nonzero values is important for several reasons. One is to find reasons why a model does not fit and to obtain hints on how to improve the model. Patterns of deviations sometimes suggest better models. For instance, when variables are ordinal, a pattern of positive and negative residuals can indicate trends neglected by the model (Agresti, 1984). Typically, however, one considers the absolute size of residuals. Haberman (1973, 1978) or Nelder (1974) suggest methods for the statistical analysis of residuals.

Because log-linear models analyze d-dimensional cross-classifications of d variables, the number of contingency table cells increases exponentially with d. Therefore, the sample size required for valid statistical testing can become very large. One method of residual analysis, correspondence analysis (CA; cf. Hill, 1974; Greenacre, 1984) exhausts information from bivariate association patterns only. As a consequence, the number of variables that can be simultaneously analyzed, given a sample size n, is larger than for log-linear analysis. CA is an analogue of principal component analysis. It analyzes the structure of relationships between variables. When applying CA, one assumes that associations beyond the first order are negligible.

Residual analysis with the goal to improve models and to determine the first-order association structure as principal component analysis conserve the basic goal of log-linear modeling. This goal is the analysis of association structures. A third reason for the application of residual analysis involves a change in perspective. *Configural frequency analysis* (CFA) (Krauth and Lienert, 1973; Lienert, 1988; von Eye, 1989) is a method of analyzing single cells in contingency tables. These cells are called *configurations*. Starting from simple log-linear models, typically the main effect model, CFA compares the expected frequency e with the observed frequency o for each cell. If, statistically,

$$o > e, \tag{1}$$

then the configuration is called a *type*. If, statistically,

$$o < e, \tag{2}$$

then the configuration is called an *antitype*. Types and antitypes describe people with certain patterns of characteristics. For people with these patterns, the model holds locally not true. Thus, CFA moves away from a focus on identifying the association structure of variables and towards a focus on individuals with specific configurations of characteristics.

Basically, CFA can refer to any log-linear model (von Eye, 1988). In most instances, however, expected frequencies are estimated under the main effect model $\log y = u_0 + \sum u_i$, where u_0 denotes the grand mean and u_i the main effect of the ith variable. This model considers only differences between frequencies of the categories of variables. The basic assumption is that all variables are independent. If CFA reveals types and antitypes under the main effect model, the assumption of independence is locally violated. (For the axiom of local independence in latent class analysis see Clogg, 1988.)

1.1. Testing for Types and Antitypes

The statistical tests used to determine whether a configuration forms a type or an antitype include the chi-square component test, the binomial test, and Lehmacher's (1981) approximate hypergeometric test.

The chi-square component test is

$$x^2 = \frac{(o - e)^2}{e}. \tag{3}$$

It is obviously closely related to Haberman's standardized residual

$$x = \frac{o - e}{e^{1/2}}. \tag{4}$$

For df $= 1$, x^2 is distributed as chi-square. Pearson's X^2 for a log-linear model is

$$X^2 = \sum_i x^2 \tag{5}$$

where i indexes all cells in the contingency table.

Since df = 1, the equation

$$x^2(\alpha) = z^2\left(\frac{\alpha}{2}\right)$$

holds, and (4) can easily be transformed into a standard normal, one-sided z-test statistic.

Nelder (1974) discusses an alternative to x that is more nearly normally distributed. This test statistic is

$$x^* = \frac{3(o^{2/3} - (\underline{e} - 1/6)^{2/3}}{2e^{1/6}}, \tag{6}$$

where $\underline{e}$ is Euler's constant 2.71828. Because of its simplicity, most users prefer the chi-square component test.

The binomial test, an exact test for types and antitypes, was suggested by Krauth and Lienert (1973). One tests for the presence of types using

$$p(o > e) = \sum_{i=o}^{n} \binom{n}{i} p^i q^{(n-i)}, \tag{7}$$

where $p = e/n$, $q = 1 - p$, and n denotes the sample size. Accordingly, one tests for the presence of antitypes using

$$p(o < e) = \sum_{i=o}^{o} \binom{n}{i} p^i q^{(n-i)}. \tag{8}$$

Because of the term $\binom{n}{i}$, the application of the binomial test is tedious.

Also, the test is lacking in power, thus leading only to the detection of very extreme types or antitypes. However, the x- and z-tests are also conservative.

Therefore, Lehmacher (1981) suggested a more powerful test of residuals based on the hypergeometrical distribution. The exact version of Lehmacher's test is sufficiently approximated by the standard normal

$$z = \frac{o - e}{s} \tag{9}$$

where

$$s^2 = e(1 + \bar{e} - e), \tag{10}$$

and

$$\bar{e} = \frac{1}{(n-1)^{d-1}} \prod_{i=1}^{d} (f_{ij} - 1), \tag{11}$$

where f_{ij} denotes the jth marginal sum of the ith variable, and d and n denote, as before, the number of variables and the sample size, respectively. Lehmacher's test is sufficiently exact for large sample sizes. For smaller samples, Küchenhoff (1986) suggested subtracting 0.5 from the numerator of (9) as a continuity correction. Notice that Lehmacher's test can be applied only if the expected frequencies are estimated under the assumption of total independence.

1.2. α Adjustment

Suppose a researcher analyzes the d variables $A, B, C, \ldots, D$. The complete cross-tabulation of these variables then has

$$t = abc \cdots d \qquad (12)$$

cells, where the lowercase letters denote the number of categories of the respective variables. CFA applies a significance test to each of the t cells. Because of this large number of tests, the critical level α must be adjusted. This adjustment prevents inflation of the experimentwise Type I error rate that is a large increase of the factual relative to the nominal α. Krauth and Lienert (1973) suggested the well-known Bonferroni adjustment. This procedure yields

$$\alpha^* = \frac{\alpha}{t}. \qquad (13)$$

A more efficient procedure has been suggested by Holm (1979). Rather than keeping the adjusted α constant for all tests, Holm suggested taking the number of tests already performed into account. The Holm-adjusted α is

$$\alpha^* = \frac{\alpha}{t - i}, \qquad i = 0, 1, \ldots, t - 1, \qquad (14)$$

where i denotes the number of tests already performed. For the first test, when $i = 0$, Holm's method yields the same adjusted α as Bonferroni's method. Beginning with the second test, however, Holm's methods gives less conservative decisions than Bonferroni's method. For the last test, $\alpha^* = \alpha$. Notice that the test statistics must be rank-ordered before application of Holm's procedure. Holm's procedure is terminated when for the first time the null hypothesis, $o = e$, cannot be rejected. Recently, even more efficient methods have been suggested by Shaffer (1986) and Holland and Copenhaver (1987, 1988; cf. Westermann and Hager, 1986).

2. Configural Frequency Analysis in Longitudinal Research

Several variations of CFA have been suggested for the analysis of longitudinal data. One goal of each of these approaches is to generate cross-tabulations containing each individual only once. Tables containing subjects more than once artificially increase the sample size. Such tables result, for instance, from simultaneously analyzing joint frequency distributions of variables observed at two points in time. Table 16.1 gives an example of such a cross-tabulation.

Table 16.1 shows a joint frequency distribution scheme for two dichotomous variables A and B, both observed at two occasions. The left-hand panel shows the cross-tabulation $A \times B$ at the first, the right-hand panel at the second occasion. Both occasions include the total sample. Thus, an analysis of this table with, for instance, X^2 statistics, would imply an artificial increase of the sample size by a factor of 2.

The usual way to handle this problem is to cross the cross-tabulation $A \times B$ at occasion 1 with the cross-tabulation at occasion 2. Thus, a $(2 \times 2) \times (2 \times 2)$ contingency table would result (cf. Bishop, Fienberg, and Holland, 1975; Knoke and Burke, 1980). Table 16.2 gives an example of such an analysis, adapted from Krauth and Lienert (1973; cf. von Eye, 1990). In this experiment, 60 subjects performed calculations. After the first and seventh series of calculations, performance in calculations (C) and flicker fusion threshold (F) were measured. Each of the variables was dichotomized at its overall median. Thus, four variables result: calculation performance after trial 1 (C_1) and calculation performance after trial 7 (C_7), flicker threshold after trial 1 (F_1) and flicker

TABLE 16.1. Frequency Distribution of Dichotomous Variables A and B Observed at Two Occasions

Variables AB	Occasion 1	Occasion 2
11	f_{111}	f_{211}
12	f_{112}	f_{212}
21	f_{121}	f_{221}
22	f_{122}	f_{222}
	n	n

TABLE 16.2.   CFA of the Variables Flicker Threshold and Performance in Calculations Observed at Two Occasions

Configurations						
C_1	C_7	F_1	F_7	o	e	$p(o = e)$
1	1	1	1	11	3.718	0.0021 T
1	1	1	2	2	5.577	0.1477
1	1	2	1	1	2.002	0.8017
1	1	2	2	0	3.003	0.0919
1	2	1	1	3	4.862	0.5466
1	2	1	2	12	7.293	0.1104
1	2	2	1	0	2.618	0.1376
1	2	2	2	4	3.927	0.8832
2	1	1	1	1	3.042	0.3706
2	1	1	2	4	4.563	0.9680
2	1	2	1	7	1.638	0.0024 T
2	1	2	2	0	2.457	0.1627
2	2	1	1	0	3.978	0.0326
2	2	1	2	6	5.967	0.8856
2	2	2	1	1	2.142	0.7274
2	2	2	2	8	3.213	0.0287

$$n = 60$$

1... = 33, 2... = 27
.1.. = 26, .2.. = 34
..1. = 39, ..2. = 21
...1 = 24, ...2 = 36

threshold after trial 7 (F_7). Together, these variables form the $2 \times 2 \times 2 \times 2$ cross-tabulation given in Table 16.2.

CFA of the data in Table 16.2 estimated expected frequencies under the main effect model of total independence of the variables C_1, C_2, F_1, and F_2. As a significance test, the binomial test given in (7) was used in its two-tailed version. Bonferroni's method was used for α adjustment. An $\alpha = 0.05/16 = 0.003125$ resulted.

Application of CFA revealed two types and no antitypes. The first type has pattern 1111. It contains subjects who perform at the first and seventh occasions better than average in calculations and have above average (i.e., lower) flicker threshold. Under the assumption of total independence of C_1, C_2, F_1, and F_2, 3.7 subjects were expected to show this pattern. However, 11 were observed. The second local contradiction of total independence occurs for configuration 2121. This type contains subjects who consistently perform below average in calculations but who also consistently show above average flicker thresholds. Seven rather than the expected 1.6 subjects displayed this pattern.

2.1. The Problem of Sample Sizes in Longitudinal CFA

The example in Table 16.2 shows how types reflecting change can result from a cross-tabulation of repeatedly observed variables. This approach is parallel to log-linear modeling of repeated observations (see Bishop et al., 1975, Chap. 7). However, both log-linear modeling and CFA face limitations in their applicability to longitudinal data. The main limitation is that the sample size required for valid analysis increases exponentially with the number of variables and occasions. For example, suppose a researcher analyzes five dichotomous variables measured at three occasions. The complete cross-tabulation would have $(2^5)^3 = 32,768$ cells.

To ensure valid goodness-of-fit or CFA tests, a minimum sample size must be met. This minimum is typically expressed in terms of a minimum required average cell frequency. For instance, to obtain a valid Pearson X^2-test, Cochran (1954) recommends an expected frequency of at least five for 80% or more of the cells. In addition, no expected frequency should be less than 1. Over the years, however, these recommendations have become less and less restrictive. In 1978, Larntz showed in simulation studies that Pearson's X^2 performs better than other X^2-tests even if e is as low as 0.5 and the sample size is no less than $n = 8$. For other tests, similar considerations apply.

In more general terms, suppose d variables with $d_1, d_2, \ldots, d_d$ states have been observed T times. Then, a complete cross-tabulation of these variables has

$$c = t^T \tag{15}$$

cells where $t = d_1 d_2 \cdots d_d$. Let n' denote the expected number of cases per cell required as a minimum for valid significance testing. Then, the smallest appropriate sample for the c cells is $n'c$. In our example, we would then conclude that we need a sample of either about 163,840 or no less than 16,384 cases for the 32,768 cells. Considerind the relatively small number of variables and occasions in this example, both sample sizes are large. To make things even worse, these minimal sample sizes are only the basis for the applicability of a statistical test. Much larger samples are necessary to obtain types and antitypes.

2.2. Approaches to Multivariate Longitudinal CFA

Several suggestions have been made concerning the reduction of the large samples needed for valid testing in multivariate longitudinal CFA (cf. von Eye and Nesselroade, 1989; von Eye, 1989). This section will briefly discuss the method of differences and the approximation of time series with orthogonal polynomials. The latter method is most efficient when the number of occasions is large. Later, a method will be presented that is most efficient when the number of variables is large.

Suppose a researcher observes a sample of subjects T times on d variables. He or she may choose to focus on subjects' change patterns across the T points in time. One aspect of change is the difference between time-adjacent observed values (cf. Games, volume one; Nesselroade and Burr, volume one). Difference scores are either

$$d = \begin{cases} + & \text{if } y_1 < y_2, \\ = & \text{if } y_1 = y_2, \\ - & \text{if } y_1 > y_2, \end{cases} \qquad (16)$$

or

$$d' = \begin{cases} + & \text{if } y_1 < y_2, \\ - & \text{if } y_1 > y_2. \end{cases} \qquad (17)$$

Difference scores are applicable only to scores observed at least at the ordinal level. If the resolution of scores is so fine that no two adjacent values are identical, the transformation given in (17) is the method of choice. For rank data or more grossly measured variables, the transformation (16) is appropriate.

The transformations (16) and (17) have two major advantages. First, they solve the problem of dependent measures. Rather than including each subject T times in the same analysis these transformations produce one characteristic vector of up-and-downs per subject. Second, the number of data points per subject drops from dT to $d(T-1)$. Thus, for d variables transformed by either (16) or (17), one obtains a cross-tabulation with

$$c = t^{(T-1)} \qquad (18)$$

cells, where $t = d_1 d_2 \cdots d_d$. Suppose, again, a researcher analyzes five variables that were observed three times. Then transformation (17) yields a cross-tabulation with $c = (2^5)^2 = 1024$ cells, a number considerably smaller than the $c = 32,768$ calculated from (15). Notice, however, that transformation (16) would result in a table with $c = (3^5)^2 = 59,049$ cells. Thus transformation (16) helps reduce the number of cells only if, before transformation, each variable has three or more states.

Transformations (16) and (17) sacrifice information on the absolute elevation of the time series. For instance, in clinical research, one might sacrifice information on the level from which a change process started. Including this information as an additional variable makes the problem as complex as before transformation. In addition, in some instances, in particular for measures derived from tests constructed using classical test theory, difference values may not be valid (see Nesselroade and Burr, volume one).

Therefore, another method for analyzing multivariate time series with CFA was suggested. This method is the approximation of each univariate time series by an orthogonal polynomial. The polynomial coefficients may then be categorized and analyzed with CFA (cf. Krauth and Lienert, 1978; Krauth, 1980, von Eye and Nesselroade, 1989).

With a general curvilinear regression approach, a series of observed values can be approximated by the polynomial

$$\hat{y} = b_0 x^0 + b_1 x^1 + b_2 x^2 + \cdots = \sum_{i=0}^{j} b_i x^i \tag{19}$$

where the x are the values of the predictor, the b_i are the estimated regression coefficients, and j is the degree of the polynomial ($0 \leq j \leq T - 2$). With longitudinal data, x denotes some value on the time axis. For example, x could represent learning trials, time passed since the beginning of a therapy, or hours spent under an experimental condition; y is the expected outcome value.

Unless $j = T - 2$, the polynomial does not necessarily hit the observed values. Instead, it smoothes the time series. As the degree of a polynomial increases, the estimated values come, on average, closer to the observed ones. Estimation is typically done using least squares methods.

Parameters of polynomials vary with their degree. If one calculates, for example, a second-degree polynomial and then realizes one needs a third-degree polynomial for an acceptable fit, a complete recalculation is necessary. It is not enough to estimate just the third-degree coefficient a.

A second disadvantage of ordinary polynomials is that the parameters are not necessarily commensurable. In other words, coefficients from different polynomials on the same data do not necessarily belong to the same scale. Therefore, any substantive interpretation is most likely invalid.

Because of these problems, CFA uses coefficients from *orthogonal polynomials*. With orthogonal polynomials, one obtains instead of (19)

$$\hat{y} = a_0\beta^0 + a_1\beta^1 + a_2\beta^2 + \cdots + a_i\beta^i = \sum_{i=0}^{j} a_i\beta^i \qquad (20)$$

where β^i denotes a polynomial of the ith degree, and $i = 0, 1, \ldots, T - 2$. For instance, to smoothe a time series with a first-order, or linear polynomial, one obtains

$$\hat{y} = a_0(k) + a_1(bx + h)$$

where the terms in parentheses represent β^0 and β^1, respectively, and k is a constant. A quadratic polynomial is given by

$$\hat{y} = a_0(k) + a_1(b_x + h) + a_2(lx^2 + mx + n)$$

where the last term in parentheses represents β^2.

Systems of polynomials as given in (20) are orthogonal on a given interval if the sum of the polynomial values for this interval is 0 across and within polynomials (see Abramowitz and Stegun, 1970; cf. von Eye and Nesselroade, 1989). Values for most polynomials can be found, for instance, in analysis of variance textbooks (e.g., Kirk, 1982).

Orthogonal polynomials have two primary advantages over ordinary polynomials. The first is that their coefficients can be interpreted in a fashion analogous to regression coefficients. Specifically, a_0 is the arithmetic mean of the observed values, a_1 is the linear regression coefficient, a_2 the quadratic regression coefficient, etc.

The second advantage is more important in the present context. Approximating time series by orthogonal polynomials can reduce the number of values for further statistical analysis. Suppose a researcher observes three dichotomous variables 10 times each. Then, application of (15) yields 1,073,741,824 cells for the cross-tabulation of these repeated observations. Even after application of the transformation given in (17), the table still has 134,217,728 cells. If the researcher approximates each of the three time series by a quadratic orthogonal polynomial, then a_0, a_1, and a_2 are the parameters analyzed with CFA. After, say, dichotomization, the researcher would have a table with 512 cells.

This example shows that approximation of time series with polynomials and subsequent analysis with CFA can lead to a dramatic reduction in size of the contingency table. The main advantage is that the problem size becomes, to a certain extent, independent of the length of the time series. (For an application of this method see von Eye and Nesselroade, 1989.) However, even for small numbers of variables, the size of the resulting cross-tabulation may still be too big for analysis with conventional sample sizes. Therefore, the following sections discuss an approach for reducing the size of the table with respect to the number of variables (von Eye, 1984).

2.3. Transformations in Multivariate Time Series

The methods for CFA analysis of longitudinal data discussed here focused on

1. Raw scores
2. Relationships between raw scores
3. Parameters of entire time series

The method suggested in the following sections focuses on the relationships between adjacent vectors of observations. While the previous approaches lead to a reduction in the size of the contingency table through decreased numbers of observations, the present approach leads to a reduction through decreased numbers of variables. The numbers of variables are reduced by comparing vectors of variables adjacent in time series.

The relationship between two vectors of measures can be investigated using coefficients of similarity or dissimilarity. Coefficients of similarity meet with the following three axioms. The first axiom is

$$A_1: \quad s_{XX} = 1,$$

where s is a coefficient reflecting the similarity between vectors X and Y. This axiom restricts the maximum value of s to 1. One obtains this maximum value if one measures the similarity of an object with itself. The second axiom is

$$A_2: \quad s_{XY} = s_{YX}.$$

This is termed the *symmetry axiom*. It indicates that if two vectors are compared with each other, the degree of similarity is the same, regardless of whether X is compared to Y or Y to X. (For a discussion of asymmetrical relationships see von Eye and Rovine, In press.) Axiom 3 is

A_3: $0 \leq s_{XY} \leq 1$.

Axiom 3 specifies the range of admissible values. A value of $s = 0$ results if there is no similarity between vectors X and Y. Increasing similarity and, thus, increasing predictability reaches its maximum value as specified in Axiom 1. Other formulations of Axiom 3 include

A_3: $0 \leq |s_{XY}| \leq 1$,

A_3: $s_{XY} \leq s_{XX}$, and

A_3: $|s_{XY}| \leq |s_{XX}|$.

In cluster analysis, in particular, a fourth axiom has been discussed. Consistent with an intuitive understanding of similarity, the fourth axiom specifies for the two pairs of vectors A and B and X and Y:

A_4: If there is one characteristic in which A and B correspond and X and Y differ, and if A and B and X and Y correspond in all other characteristics, then

$$s_{XY} < s_{AB}.$$

Axiom 4 specifies that the similarity between two vectors decreases with the number of corresponding elements. There is no similarity coefficient that does not meet with Axiom 4.

For coefficients of dissimilarity or distance, similar axioms have been discussed. However, only the following two axioms are nonredundant (Young, 1987).

A_1: $s_{XY} = 0$ if and only if $X = Y$.

This axiom specifies that the distance between two identical points is always zero. A single point can never occupy more than one location. Two different points can never have a zero distance.

A_2: $s_{XY} \leq s_{XZ} + s_{YZ}$.

This axiom represents the triangular inequality. It specifies that the sum of two edges of a triangle can never by less than the remaining edge. Symmetry and nonnegativity can be derived from distance Axioms 1 and 2.

There are many coefficients of similarity and dissimilarity or distance that meet with these axioms. Large coefficients indicate strong similarity, large distance, or strong dissimilarity. For quantitative scales, Pearson's product-moment correlation coefficient, Euclidean distance, and Mahalanobis's generalized distance are among the most widely used measures. The selection of measures is not straightforward. The problem is that there rarely is a mapping between assumptions of social science theories and characteristics of the coefficients. This chapter selects the product-moment correlation coefficient as a measure of similarity and the Euclidean distance as a measure of distance.

2.4. CFA of Multivariate Time Series Based on Correlations between Adjacent Vectors of Measures

Suppose a researcher observes n subjects T times on d variables. The data from these observations form an $n \times T \times d$ data box (Cattell, 1988). Each "slice" of this data box contains the T data vectors of an individual. To reduce the size of the cross-tabulation for CFA, one can perform the following transformation. One correlates each row of the $T \times d$ matrix of measures observed per person with the next row. As a result, one obtains a vector with $T - 1$ correlations for the matrix of $T \times d$ measures.

For quantitative scales one can use the product-moment correlation. For centered scales this coefficient is

$$r = ((X^{T}X)^{1/2}(Y^{T}Y)^{1/2})^{-1}(X^{T}Y) \tag{21}$$

where X and Y denote the time-adjacent vectors. In the present context, (21) measures the autocorrelation for a person. Usually, (21) measures the correlation between two independent random vectors. The following are important properties of r.

1. r can be computed only if the variances for both X and Y are different from zero. In other words, the values observed across variables for a person at any given occasion must not be all the same. If this requirement is met, the admissible range of values of r is from -1 to $+1$.
2. $r_{XY} = r_{YX}$ (symmetry).
3. $r_{XX} = 1$.
4. If the two variables are dichotomous, one obtains $r_{xy} = \{-1, +1\}$. If $r = +1$, the product $(x_1 - y_1)(x_2 - y_2)$ is positive; otherwise the product is negative.

It is well known that r measures the degree of linear relationship between two vectors. If the relationship between X and Y is nonlinear, r will not reflect that relationship. (Notice, however, that r-like measures can be calculated for nonlinear relationships also.) Equation (21) is valid even if the scales of measurement of two variables are different. In the present context, however, they will always be commensurate, because the same variables are repeatedly observed and the scales remain the same. The measure r is independent of the means and standard deviations of X and Y. Therefore, based on this characteristic, one can more precisely specify what r measures. The product-moment correlation r measures the similarity of shapes of the two vectors of measures. In other words, r measures the degree to which the up-and-downs of the measures in X and Y correspond. The more accurately one can predict the up-and-down of one vector from a linear transformation of the other, the larger r.

In certain applications, r is hard to interpret. For instance, the value $r = 0$ allows one to conclude stochastic independence of X and Y only if the d pairs $x_i - y_i$ are realizations of the same binomially distributed random vector. Also, the angular interpretation of r is possible only if X and Y have the same length.

For two reasons, however, r is a suitable measure in the present applications. First, we are interested in the similarity of the shapes of X and Y. (The next section discusses the spatial distance of X and Y.) Second, we assume the problems with the interpretation of r do not apply here, because we assume the repeatedly observed variables are constant in their characteristics. Therefore, the application of r is valid.

Table 16.3 depicts how T adjacent times series can be transformed into

TABLE 16.3. Computation of Autocorrelations for CFA of Multivariate Time Series

Subject	Vectors at point in time		Correlation
	$t - 1$	t	
1	$\{x_{111}, x_{112}, x_{113}, \ldots, x_{11d}\}$	$\{x_{121}, x_{122}, x_{123}, \ldots, x_{12d}\}$	r_{11}
	$\{x_{121}, x_{122}, x_{123}, \ldots, x_{12d}\}$	$\{x_{131}, x_{132}, x_{133}, \ldots, x_{13d}\}$	r_{12}
	$\vdots$	$\vdots$	$\vdots$
	$\{x_{1m1}, x_{1m2}, x_{1m3}, \ldots, x_{1md}\}$	$\{x_{1T1}, x_{1T2}, x_{1T3}, \ldots, x_{1Td}\}^a$	r_{1m}
$\vdots$	$\vdots$	$\vdots$	$\vdots$
n	$\{x_{n11}, x_{n12}, x_{n13}, \ldots, x_{n1d}\}$	$\{x_{n21}, x_{n22}, x_{n23}, \ldots, x_{n2d}\}$	r_{n1}
	$\{x_{n21}, x_{n22}, x_{n23}, \ldots x_{n2d}\}$	$\{x_{n31}, x_{n32}, x_{n33}, \ldots, x_{n3d}\}$	r_{n2}
	$\vdots$	$\vdots$	$\vdots$
	$\{x_{nm1}, x_{nm2}, x_{m3}, \ldots, x_{nmd}\}$	$\{x_{nT1}, x_{nT2}, x_{nT3}, \ldots, x_{nTd}\}^a$	r_{nm}

[a] $m = T - 1$.

$T - 1$ correlations. The first row shows that r_1 results from correlating the vectors from occasions 1 and 2. The first subscript in the vectors indexes the subjects $(1, \ldots, n)$. The second subscript indexes occasions $(1, \ldots, T;$ with $m = T - 1)$, and the third subscript indexes variables $(1, \ldots, d)$; r_2 results from correlating the vectors from occasions 2 and 3, and so forth.

For further analysis with CFA or log-linear modeling, these correlations must be categorized. The resulting contingency table has $c = t$ cells, where t is defined as in (15) or (18). For example, if a researcher observes 15 variables four times, application of the transformation depicted in Table 16.3 yields three correlations. After dichotomization, the table for CFA has only $c = 2 \cdot 2 \cdot 2 = 8$ cells. The same number of cells results for any number of variables observed four times. The complete cross-tabulation of the untransformed variables has 1.153×10^{12} cells.

2.5. CFA of Multivariate Time Series Based on Distances between Adjacent Vectors of Measures

Distance measures may be used in place of correlation coefficients. Thus, this section follows the same procedure as the last one. The only difference is that the Euclidean distance is the measure for the transformation of the neighbored vectors X and Y. A formula for the Euclidean distance is

$$l = \left(\sum_{i=1}^{d} (x_i - y_i)^2 \right)^{1/2} \tag{22}$$

where x_i and y_i denote the elements of the two adjacent vectors. The Euclidean distance is a special case of the general Minkowski metric (L-metric), which is

$$L = \left(\sum_{i=1}^{d} |x_i - y_i|^u \right)^{1/u}, \qquad u \geq 1. \tag{23}$$

As one can see from (23), $u = 2$ yields the Euclidean distance. For $u = 1$ one obtains the city-block metric

$$l = \sum_{i=1}^{d} |x_i - y_i|. \tag{24}$$

The properties of the Euclidean distance are well known: l measures the spatial distance between two points in a d-dimensional space. If this space has only two or three dimensions, l corresponds with our intuitive understanding of distance.

The distance l is defined even if all elements of the vectors X and Y are numerically identical. The Euclidean metric defines a d-dimensional space with orthogonal axes. Therefore, the computation of l is meaningful only if the d variables are independent. If some of the variables are dependent, they obtain a weight greater than the one of the independent variables. The relative weight of variables depends, therefore, on their intercorrelations.

Implicit weighting also results from differences in the variables' standard deviations. Here, the weighting is proportional to the differences in standard deviations. A third cause of unequal weighting of variables is incommensurability. If one variable's unit is miles and the other's unit is millimeters, the second variable obtains a much larger weight than the first. Here, weighting is proportional to the ratio of the variables' units.

Principal component analysis has been suggested as a method to eliminate weighting caused by scale intercorrelations. This method, however, assumes that the variation in the data space can be exhausted by first-order correlations. Therefore, it is counterindicated as a step before CFA. If one is interested in the pattern of types and antitypes beyond first-order scale intercorrelations, one may perform second-order CFA. This is a variant of CFA that estimates expected frequencies under consideration of first-order associations. Thus, types and antitypes emerge only if there are associations beyond the first order (von Eye, 1989).

If the standard deviations differ, a standardization via z-transformation may be considered. This transformation also makes scales commensurable, a must in the application of l.

A transformation of raw data vectors using l is recommended when the measures can be interpreted as a profile, and the level of the profile is of interest. The distance l_{xy} provides information on the location of profiles in the data space. In contrast to the correlation coefficient, l is not sensitive to the up-and-downs of the measures within a vector. The only exception concerns differences in standard deviations reflected by the up-and downs. Therefore, when both r and l are correctly applied, high correlations between X and Y can occur regardless of the size of the distance, and vice versa. Figure 16.1 illustrates this relationship. It depicts the three profiles A, B, and C, which display the following relationships.

1. $r_{AC} = -1$ and $l_{AC} = $ small.

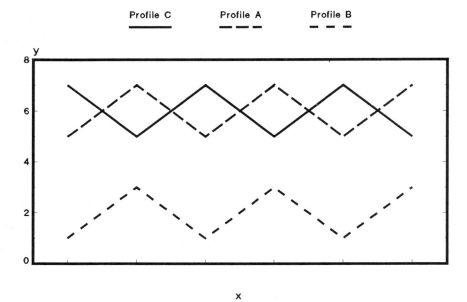

FIGURE 16.1.   Comparison of correlation and distance.

2. $r_{AA} = +1$ and $l_{AA} = 0$.
3. $r_{AB} = +1$ and $l_{AB} = $ large.
4. $r_{BC} = -1$ and $l_{BC} = $ large.

In a manner analogous to Table 16.3, Table 16.4 shows how distances can be calculated from raw data. The top row shows that the first distance results from the comparison of the first two vectors with each other. The second distance results from the comparison of the second with the third vector, and so on.

For analysis with CFA, the distances must be categorized. There are two approaches to categorizing distances (von Eye, 1984). The first is to categorize the distances as they result from the transformation depicted in Table 16.4. This approach does not consider the direction of change. The second approach considers the directions of the shifts in location. A sign for the distances can be defined as follows:

$$l = \begin{cases} -1 & \text{if } \bar{x} < \bar{y} \\ 1 & \text{else} \end{cases}, \tag{25}$$

TABLE 16.4. Computation of Autodistances for CFA of Multivariate Time Series

Subject	Vectors at point in time		Distance
	$t-1$	t	
1	$\{x_{111}, x_{112}, x_{113}, \ldots, x_{11d}\}$	$\{x_{121}, x_{122}, x_{123}, \ldots, x_{12d}\}$	l_{11}
	$\{x_{121}, x_{122}, x_{123}, \ldots, x_{12d}\}$	$\{x_{131}, x_{132}, x_{133}, \ldots, x_{13d}\}$	l_{12}
	$\vdots$	$\vdots$	$\vdots$
	$\{x_{1m1}, x_{1m2}, x_{1m3}, \ldots, x_{1md}\}$	$\{x_{1T1}, x_{1T2}, x_{1T3}, \ldots, x_{1Td}\}^a$	ll_m
$\vdots$	$\vdots$	$\vdots$	$\vdots$
n	$\{x_{n11}, x_{n12}, x_{n13}, \ldots, x_{n1d}\}$	$\{x_{n21}, x_{n22}, x_{n23}, \ldots, x_{n2d}\}$	l_{n1}
	$\{x_{n21}, x_{n22}, x_{n23}, \ldots, x_{n2d}\}$	$\{x_{n31}, x_{n32}, x_{n33}, \ldots, x_{n3d}\}$	l_{n2}
	$\vdots$	$\vdots$	$\vdots$
	$\{x_{nm1}, x_{nm2}, x_{nm3}, \ldots, x_{nmd}\}$	$\{x_{nT1}, x_{nT2}, x_{nT3}, \ldots, x_{nTd}\}^a$	l_{nm}

[a] $m = T - 1$.

where $\bar{x}$ and $\bar{y}$ are the arithmetic means of the two vectors. Without signs, types and antitypes of shifts in location indicate simply the size of the shifts. There is no discrimination between large upward and large downward shifts. Thus, for example, clinical patients whose psychiatric condition worsens cannot be distinguished from patients who get better. Therefore, the assignment of signs is always recommended.

3. A Data Example: Changes in Mood

This section applies CFA to multivariate repeated measures of mood (von Eye, 1984). In a memory experiment subjects had to learn and recall two short texts. Before and after processing each text; the mood variables anxiety, arousal, and fatigue were measured with a questionnaire. Thus, a series of four measures per mood item resulted. The range of possible values per mood variable is $0 \le x \le 36$. Fatigue is inversely scaled. Thus, high scores indicate lack of fatigue. A total of 148 subjects participated in the experiment: 49 young adults (18–32 years old), 49 middle-aged adults (33–55 years old), and 50 older adults (above 55 years old).

Data analysis will use both correlation and distance transformations of raw scores. First, the four vectors of three measures each will be correlated. A vector with three correlations per person will then result and, after categorization, will be analyzed with CFA. Second, the

Euclidean distances of the measures will be calculated and analyzed with CFA.

3.1. CFA of Autocorrelations of Mood

With the scheme in Table 16.3, three correlations per person were computed. The distribution of these correlations across subjects is bimodal. The first mode is located at $r = 0.99$, the second at $r = -0.80$. In addition, the distribution is asymmetrical. There are far more positive than negative correlations. Therefore, correlations were not dichotomized at their median, located at $r = 0.90$. Rather, to discriminate the high correlations from the lower or negative ones, the cutoff was set at $r = 0.50$. At $r = 0.50$, the "valley" between the two modes of the bimodal distribution reaches its lowest point.

The three dichotomous correlations were crossed to form a $2 \times 2 \times 2$ contingency table. Table 16.5 gives the cross-tabulation and summarizes results of the CFA. The a priori α was 0.05, the critical α was determined using Holm's (1979) method. A starting value of $\alpha = 0.00625$ resulted. As a significance test, Lehmacher's approximate test with continuity correction (Küchenhoff, 1986) was used. Expected frequencies were estimated under the main effect model of total independence of the three autocorrelations.

The first three columns of Table 16.5 show the autocorrelation patterns that result from the dichotomization of correlations at $r = 0.50$. For instance, the first 1 in pattern 111 indicates a very strong autocorrelation

TABLE 16.5. CFA of the Autocorrelations of Three Mood Variables

Correlations 123	o	e	$\lvert z \rvert$	$p(z)$	$p_{\text{crit.}}$
111	65	56.41	2.61	0.004 T	0.006
112	12	13.16	0.27	0.393	0.017
121	31	38.46	2.33	0.010	0.008
122	9	8.97	0.21	0.418	0.025
211	8	14.95	2.55	0.005 A	0.007
212	3	3.49	0.008	0.497	0.050
221	16	10.19	2.23	0.013	0.010
222	4	2.38	0.80	0.213	0.013

T = Type; A = Antitype

between the measures before and after the first trial of the memory experiment. The second 1 indicates a very strong autocorrelation between the mood measures after the first and before the second trial. The last 1 indicates a very strong autocorrelation between the mood measures before and after the second trial.

CFA revealed one type and one antitype. The type has pattern 111. This pattern suggests that the shape of mood measures of almost half of the subjects was unaffected by the memory experiment. This type contains subjects with stable mood patterns. Notice that this result concerns only the up-and-downs of the measures in each vector. Figure 16.1 shows that shifts in location may have occurred but autocorrelation would not reflect these shifts. For the subjects with pattern 111, however, these shifts, if they occurred, must be parallel.

The antitype has correlation pattern 211. The 2 in this pattern indicates that the correlation between the mood measures before and after the first experimental trial is low or negative. After the first trial, the shape of mood indicators is stable. In other words, the antitype contains subjects whose mood changed during the first experimental trial, but remained stable for the rest of the experiment. Significantly fewer subjects than expected under the null hypothesis of total independence of the autocorrelations displayed this pattern.

3.2. CFA of Shifts in Mood

With the scheme in Table 16.4, three distances per person were calculated. In addition, to identify the direction of shifts, distances were assigned signs using equation (25). The distribution of distances was less skewed than the distribution of correlations. Therefore, the dichotomization of distances used the origin 0 as a natural curoff. Shifts toward numerically greater values were given a 1, and decreases were given a 2. Table 16.6 gives the cross-tabulation of the dichotomous distance scores and summarizes results of the CFA. The same methods were applied as in Table 16.5.

The first three columns in Table 16.6 show the distance patterns that resulted from dichotomization of distances at zero. The first 1 of, for instance, configuration 111 indicates that mood values were higher after than before the first trial of the memory experiment. Subjects who show this pattern report, after the first trial, increased anxiety and arousal and decreased fatigue. The second and the third 1 in this configuration indicate that the increase in anxiety and arousal and the decrease in

TABLE 16.6.   CFA of the Autodistances of Three Mood Variables

Distances 123	o	e	$|z|$	$p(z)$	p_{crit}
111	17	26.25	2.67	0.004 A	0.010
112	18	17.40	0.03	0.487	0.025
121	38	28.87	3.91	$5 \cdot 10^{-5}$ T	0.007
122	12	16.48	1.36	0.09	0.013
211	16	19.45	0.97	0.167	0.017
212	25	12.90	4.23	$1 \cdot 10^{-5}$ T	0.006
221	18	18.43	0.02	0.491	0.050
222	4	12.22	2.86	0.002 A	0.008

T = Type; A = Antitype

fatigue continued in the pause between the two trials and during the second trial.

CFA revealed two types and two antitypes. The first type has distance pattern 121, indicating that after the first trial 38 individuals, which is the largest group of subjects, showed higher anxiety and arousal scores but lower fatigue scores. After the pause, these subjects displayed an inverted trend: Anxiety and arousal scores were lower and fatigue higher. The second experimental trial increased mood values again. The second largest group of subjects, 25 subjects, or about twice as many as expected showed pattern 212. This trend is exactly opposite that of the first type. These subjects' mood values decreased during the experimental trials and increased during the pause.

The two antitypes have patterns 111 and 222. These antitypes indicate that it is highly unlikely that subjects display consistently increasing or decreasing mood values. One may, therefore, conclude that the experiment not affecting the mood level of these subjects is highly unlikely.

4.   Discussion

This chapter introduces the reader to CFA of multivariate repeated measures. One of the major problems of the analysis of repeated observations with log-linear models and CFA is that the number of contingency table cells increases exponentially with the number of occasions and variables. This chapter discusses three approaches to reducing the size of tables. The first approach is to analyze change via difference rather than raw scores. The gain in terms of decreased number of cells depends upon the number of occasions. Only for short time series is the gain considerable. This approach sacrifices information on the elevation of the time series.

The second approach is most efficient when there are many observation points. This approach approximates time series using orthogonal polynomials. The polynomial coefficients are categorized and then analyzed with CFA. This analysis considers both level and change information. It decomposes change information into linear, quadratic, or higher-order trends.

This chapter focuses on a third approach, most efficient when the number of variables is large. Here, the data vectors obtained at each occasion are related to their neighboring vectors. Methods to relate data vectors to each other include correlation and distance measures. The $d \times T$ data points per person are reduced to $T - 1$ correlation or distance measures per person.

There are several ways to modify and extend this approach. The first concerns the type of relationships between the vectors of repeated observations. There are literally hundreds of coefficients of similarity and dissimilarity, of information transmitted, or of association. All allow one to posit theoretical relationships in a custom tailored fashion. However, the selection of an appropriate coefficient remains a still unresolved problem. The following two criteria may be used for the selection (cf. Liebetrau, 1983).

1. *Scale level of data.* When data are nominal, one is confined to coefficients of similarity, dissimilarity, association, or information transmitted. Distances between nominal categories are not defined. For ordinal variables, distances are not defined either. However, one can use the information how many ranks observed behaviors are apart. Interval and ratio scales pose even fewer restrictions.

2. *Symmetry versus asymmetry.* As was pointed out in connection with similarity Axiom 2, correlations and distances typically are symmetrical. If the researcher is making predictions from one variable to another or from a state of a first variable to a state of a second, then asymmetrical coefficients as Goodman and Kruskal's (1954) λ, Somers' (1962) d, or the information-theoretic constriction coefficient (Attneave, 1959) might be more useful (cf. von Eye and Rovine, In press). Some of these

coefficients express the degree of asymmetric similarity in terms of the variance of one variable rather than in terms of the total variance.

In many instances variables in a repeated observation study differ in their characteristics. For instance, some variables may be nominal, others, ordinal. As long as there are two or more variables of each type, one can use the coefficient appropriate for each type. For relationships between variables from different types, the coefficients for the lowest scale level are recommended. As a result, one obtains $g(T - 1)$ measures per individual, where g is the number of measures one calculates per comparison. After categorization, one forms a composite variable from the g coefficients. This variable contains all possible configurations of all categorized coefficients in a comparison. For instance, if a researcher calculates one product moment and two rank correlation coefficients, the resulting composite variable has the states $111, 112, \ldots, 222$. To perform CFA, one crosses the $T - 1$ composite variables.

As we explained using Figure 16.1, similarity and distance coefficients are independent of each other. The same may apply for other coefficients. Therefore, another facet of the approach presented here uses both types of coefficients in the same analysis. Again, one must form composite variables before applying CFA. Resulting types and antitypes describe subjects who display a particular change pattern of, for example, correlation-distance combinations.

The present chapter discusses transformations based on comparisons of adjacent profiles. One may also compare each profile with an anchor or reference profile (cf. ideal point analysis as discussed by Wood, this volume). For instance, one could use as anchors the typical profile of a schizophrenic or the profile measured before a therapy. Training programs might work toward reducing the distance from actual to ideal or goal profiles.

The present chapter presents CFA based on correlation or distance coefficients as a method used mainly for the search of types and antitypes. However, other strategies of CFA may be considered also. For instance, the method can be used to check whether the three age groups display different mood change patterns. To do multiple group comparisons one generates a table with t cells for each of the age groups. The contingency table resulting for analysis with CFA or log-linear modeling has $c = gt$ cells, where g denotes the number of groups. For the data presented here, three-sample CFA revealed no age differences in change patterns of correlations nor distances (Krauth and Lienert, 1973; von Eye, 1990).

References

Abramowitz, M., and Stegun, I. A. (Eds.) (1970). *Handbook of mathematical functions.* New York: Dover.

Agresti, A. (1984). *Analysis of ordinal categorical data.* New York: Wiley.

Attneave, F. (1959). *Applications of information theory to psychology.* New York: Holt.

Bishop, Y. M. M., Fienberg, S. E., and Holland, P. W. (1975). *Discrete multivariate analysis. Theory and practice.* Cambridge, Mass.: MIT Press.

Cattell, R. B. (1988). "The data box: its ordering of total resources in terms of possible relational systems." In J. R. Nesselroade and R. B. Cattell (Eds.), *Handbook of multivariate experimental psychology,* 2nd ed. (pp. 69–130). New York: Plenum.

Clogg, C. C. (1988). "Latent class models for measuring." In R. Langeheine and J. Rost (Eds.), *Latent trait and latent class models.* New York: Plenum.

Cochran, W. G. (1954). "Some methods for strengthening the common χ^2 tests," *Biometrics.* **10,** 417–451.

Goodman, L. A., and Kruskal, W. H. (1954). "Measures of associations for cross-classifications," *Journal of the American Statistical Association.* **49,** 732–764.

Greenacre, M. J. (1984). *Theory and applications of correspondence analysis.* New York: Academic Press.

Haberman, S. (1973). "The analysis of residuals in cross-classified tables," *Biometrics.* **29,** 205–220.

Haberman, S. (1978). *Analysis of qualitative data,* Vol. 1: *Introductory topics.* New York: Academic Press.

Hill, M. O. (1974). "Correspondence analysis: a neglected multivariate method," *Applied Statistics.* **23,** 340–354.

Holland, B. S., and Copenhaver, D. M. (1987). "An improved sequentially rejective Bonferroni test procedure," *Biometrics.* **43,** 417–423.

Holland, B. S., and Copenhaver, D. M. (1988). "Improved Bonferroni type multiple testing procedures," *Psychological Bulletin.* **104,** 145–149.

Holm, S. (1979). "A simple sequentially rejective multiple test procedure," *Scandinavian Journal of Statistics.* **6,** 65–70.

Kirk, R. E. (1982). *Experimental design,* 2nd ed. Belmont, Calif.: Wadsworth.

Knoke, D., and Burke, P. J. (1980). *Log-linear models.* Beverly Hills: Sage.

Krauth, J. (1980). "Nonparametric analysis of response curves," *Biometrical Journal.* **30,** 59–67.

Krauth, J., and Lienert, G. A. (1973). *KFA. Die Konfigurationsfrequenzanalyse und ihre Anwendung in Psychologie und Medizin.* Frieburg, W. Germany: Alber.

Krauth, J., and Lienert, G. A. (1978). "Nonparametric two-sample comparison of learning curves based on orthogonal polynomials," *Psychological Research.* **40,** 159–171.

Küchenhoff, H. (1986). "A note on a continuity correction for testing in three-dimensional configural frequency analysis," *Biometrical Journal.* **28,** 465–468.

Larntz, K. (1978). "Small sample comparisons of exact levels for chi-squared goodness-of-fit statistics," *Journal of the American Statistical Association.* **73,** 253–263.

Lehmacher, W. (1981). "A more powerful simultaneous test procedure in configural frequency analysis," *Biometrical Journal.* **23,** 429–436.

Liebetrau, A. M. (1983). *Measures of association.* Beverly Hills, Calif.: Sage.

Lienert, G. A. (Ed.) (1988). *Angewandte Konfigurationsfrequenzanalyse.* Frankfurt, W. Germany: Athenäum.

Nelder, J. A. (1974). "Log-linear models for contingency tables: A generalization of classical least squares," *Applied Statistics.* **23,** 323–329.

Shaffer, J. P. (1986). "Modified sequentially rejective multiple test procedures," *Journal of the American Statistical Association.* **81,** 826–831.

Somers, R. H. (1962). "A new asymmetric measure of association for ordinal variables," *American Sociological Review.* **27,** 799–811.

von Eye, A. (1984). "Konfigurationsanalytische Typisierung multivariater Verlaufskurven," *Psychologische Beiträge.* **26,** 37–51.

van Eye, A. (1987). "The general linear model as a framework for models in configural frequency analysis," *Biometrical Journal.* **30,** 59–67.

von Eye, A. 1990. *Introduction to configural frequency analysis. The search for types and antitypes in cross-classifications.* Cambridge: Cambridge University Press.

von Eye, A., and Nesselroade, J. R. (In press). "Types of change: Application of configural frequency analysis to repeated observations in developmental research," *Experimental Aging Research.*

von Eye, A., and Rovine, M. J. (In press). "On concepts and measures of asymmetric similarity."

Young, F. W. (1987). *Multidimensional scaling: history, theory, and applications.* Hillsdale, N.J.: Lawrence Erlbaum.

Westermann, R., and Hager, W. (1986). "Error probabilities in educational and psychological research," *Journal of Educational Statistics.* **11,** 117–146.

Author Index

Subject Index

G

Gain scores, 82, 84–87, 103, 106
Gamma distribution mixtures, *484, 504*
General Linear Model, 44, 50
GLIM, *285*
Gompertz model, *266, 293–294, 276*
Goodness-of-fit indices, 175, 179–180, 183, 187, 194, 214, *307, 552*
Goodness-of-fit statistics, *419, 448–450, 454–455, 457, 467*
Goodness-of-fit tests, 140, *472–473, 475, 482–483, 485, 489, 491–492, 503, 505–506, 515, 523–524*
Greenhouse–Geisser (*ε*), 100
Growth curves, 60, 137, 157, 175, 191, 196, 206–207, *290, 296, 298–303, 306–307, 309, 314, 395, 397*
fitted, *289–290, 293, 304, 307, 313–314*
latent growth model (LGM), 61, 195
prediction, *293, 299, 309, 313*
Growth models, 61, 209, *291, 314*

H

Hannan–Quinn criterion (HQ), *364*
Hazard rate models, 21–23, *260–261, 263–266, 268–269, 271–274, 276–279, 282, 284*
Histograms, fitted, *491, 498*
Homogeneity, 84, 86, 91, 106–107
Hotdeck procedures, 45, 56
Human growth development, *290–315*
Hypergeometric distribution, 56, *548*
Hyperparameters, *292, 299, 301, 304*

I

IBM Job Control Language (JCL), *466*
Identification, *451, 458, 460–461, 471, 482–485*
Imputation methods (substitution), 35, 37, 42, 69
Indicator variable, 41–43, 51, *413–414, 421, 461, 467, 471, 479–480, see also* Variables
Inductive-hypothetico-deductive spiral, 6
International Mathematical and Statistical Libraries (IMSL), *460–462, 466*

Intervention effects, 4
Intraindividual relationship, *352–353, 382*
Invariance of parameters, 156, 210

J

Jackson Personality Inventory, 37, 39
Jenns model, *299, 303, 306*
Jump process, *260, see also* Event histories; Failure time processes

K

Kernel estimation, 240

L

LabVIEW, *330, 346*
Lags, *304, 383, 399*
lagged effects (associations), *426-427, 430, 432–433, 439*
time-lagged relationships, *382*
LAT, *446, 460, 506*
Latent class analysis (LCA), *444–446, 458–459, 465, 467, 495–497, 505–506, 547*
Latent classes, *445–446, 452–453, 455, 459, 467, 474, 477–480, 491, 496, 503*
Latent class models, *444–446, 456, 459, 462, 491, 494, 496, 505–506*
Latent class probabilities, *446, 477, 481–482, 486, 490, 495*
Latent dimensions, 230
Latent factors, 127
Latent moderator variable, *481, 494, 496*
Latent trait models, *459*
Latent variables, 5–6, 23, 137, 139, 140, 155–156, 167, 206, *401, 445, 450, 474–475, 477, 479, see also* Variables
categorical (discrete), *459, 479*
Latent variable equations, 51
Lawley's selection theorem, 130
LCABIN, *505–506*
LCALIN, *506*
Least squares, *303, 554*